W9-ALU-246

Protein Methods

Protein Methods

DANIEL M. BOLLAG
STUART J. EDELSTEIN

Department of Biochemistry
University of Geneva
Geneva, Switzerland

A JOHN WILEY & SONS, INC., PUBLICATION
New York • Chichester • Brisbane • Toronto • Singapore

Address all Inquiries to the Publisher
Wiley-Liss, Inc. 605 Third Avenue, New York, NY 10158-0012

Printed in the United States of America.

First Printing, January 1991
Second Printing, October 1991

Library of Congress Cataloging-in-Publication Data

Edelstein, Stuart J.
Protein methods / Stuart J. Edelstein, Daniel M. Bollag.
p. cm.
Includes bibliographical references and index.
ISBN 0-471-56871-6
1. Proteins--Purification. 2. Proteins--Research--Methodology
I. Bollag, Daniel M. II. Title.
[DNLM: 1. Electrophoresis, Agar Gel--methods--laboratory manuals.
2. Electrophoresis, Gel, Two-Dimensional--methods--laboratory manuals. 3. Immunoblotting--methods--laboratory manuals.
4. Proteins--analysis--laboratory manuals. 5. Proteins--isolation & purification--laboratory manuals. QU 25 E21p]
QP551.E23 1990
574.19'245--dc20
DNLM/DLC
for Library of Congress

90-13088
CIP

Contents

Preface

The revolution in genetic engineering has benefitted greatly from the fact that the behavior of DNA fragments created by restriction enzymes is largely independent of the fragments' precise compositions. Many properties of these fragments therefore depend essentially on their size only. This feature of DNA has permitted various generalized methods to be developed and summarized in extremely useful laboratory manuals, notably *Molecular Cloning* by Maniatis, Fritsch, and Sambrook.

Proteins, from many points of view, have more complicated personalities than DNA fragments. As a result, it has been difficult to design manuals for laboratory methods that can be applied to proteins in general. However, for routine methods concerning the elementary operations of extraction and concentration determination, as well as for certain widely used techniques involving gel electrophoresis, the distinctive properties of different proteins are not of primary importance. Hence, these methods can be applied directly to virtually all proteins.

Our goal in preparing this book has been to assemble the most general methods for protein research and to present them in practical detail in order to provide investigators with all of the information needed to perform these procedures in the laboratory. In addition, we have organized the material in the form of standardized laboratory protocols to facilitate its utilization. This orientation has led us to delve only minimally into the theory of the methods or into the basic concepts of protein biochemistry, but other sources are available for these topics.

Several guidelines have served us in designing the nature of the topics and the manner in which they are presented. The most common theme of protein analysis at the current time is gel electrophoresis, and this serves as the heart of *Protein Methods* (Chapters 5, 6, and 7). The extension of gel electrophoresis involving electroblotting and immunochemical detection of electroblotted proteins has led to some of the most impressive advances in protein analysis in recent years, and these techniques are treated in Chapter 8. Protein analysis by electrophoretic methods requires that the protein sample be extracted from its native cellular environment (Chapter 2), that the *in vitro* protein concentration be estimated (Chapter 3), and sometimes that the sample be concentrated prior to analysis (Chapter 4). Some fundamental principles for handling of proteins are outlined in Chapter 1.

In many cases, the descriptions presented are derived from procedures that have been used and refined in our own laboratory. However, their origins are multiple and are not always accurately traceable. We sincerely apologize to any uncited investigators who may be responsible for key developments. Overall, we have attempted to provide sufficient references to the original literature to enable each researcher to expand his or her knowledge of the subject and to develop specialized refinements.

We have submitted the material presented here to various colleagues for their critical evaluation, and we particularly wish to express our appreciation to Christine

Tachibana and Gideon Bollag who provided valuable comments during the development of the manuscript, as well as to Clement Bordier for his guidance on the sections concerning detergents. We thank Isabelle Tornare and Ann-Marie Paunier Doret in our laboratory for their participation in many of the developments that led to the final protocols. The skillful contributions of F. Pillonel in the preparation of the figures are gratefully acknowledged. For the weak points or errors that may remain, we take full responsibility.

A methods book of this type can always be improved by feedback from users, and we are eager to hear from those who use this book and wish to share with us their comments, criticisms, and suggestions for additions or deletions.

We are most appreciative of the people at the Wiley-Liss Division of John Wiley & Sons, Inc. who helped with this project, starting with Peter Brown, with whom it was conceived, and including Eric Swanson, Eileen Cudlipp, Frederick Siebenmann, and Sonny Fritz. Gary Giulian, Henry Reeves, and Kathleen Dannelly kindly supplied figures used in Chapters 5 and 7.

We also thank our family, friends, and colleagues upon whom we imposed for reading various sections, providing advice, and tolerating our preoccupations during the preparation of this book.

Chapter 1

Preparation for Protein Isolation

I. Introduction

This book is devoted to laboratory techniques for the analysis of proteins. Proteins are an extremely heterogeneous class of biological macromolecules. They are often unstable when not in their native environment, which in itself varies considerably among cell compartments and extracellular fluids. Of the many types of proteins, we can distinguish between those that are soluble or membrane-bound, those with catalytic or purely structural roles, and those with various post-translational modifications.

Each protein may have specific requirements once extracted from its normal biological milieu. If these requirements are not satisfied, the protein can rapidly lose its distinguishing characteristics and an already limited lifetime may be drastically reduced. Thus, determination of these requirements has often been a major hurdle in protein characterization. In some cases, the difficulty has been to stabilize the protein against external proteolysis, while in other cases the problem has been to maintain enzyme activity. Solutions to these problems are highly individual. Nonetheless, some fundamental parameters must be considered by anyone studying proteins. In this chapter, we discuss a number of these parameters and attempt to provide general guidelines or sources of information for laboratory work with proteins.

II. Buffers

A. Buffer Characteristics

- A buffer is used when working with proteins to resist changes in the hydrogen ion concentration (pH) of the protein solution. The selection of an appropriate buffer is important in order to maintain the protein at the desired pH and to ensure reproducible experimental results. A rudimentary description of key concepts behind buffering, such as pH and pK_a, can be found in the Calbiochem "Buffers" booklet and in Stryer (1988, pp. 41-42).

- There are eight important characteristics to consider when selecting a buffer (adapted from Scopes, 1982):
  1. pK_a value (see Table 1.1)
  2. pK_a variation with temperature
  3. pK_a variation upon dilution
  4. Solubility
  5. Interaction with other components (such as metal ions and enzymes)
  6. Expense
  7. UV absorbance
  8. Permeability through biological membranes

- Some General Observations

  1. Ideally, different buffers with a similar pK_a should be tested to determine whether there are undesired interactions between a certain buffer and the protein under investigation (Blanchard, 1984).

  2. Once a buffer is chosen, it is best to work at the lowest reasonable concentration to avoid nonspecific ionic strength effects. A 50mM buffer is a good starting point.

  3. The useful buffering range diminishes significantly beyond 1 pH unit on either side of the pK_a. Note that many enzymes are irreversibly denatured at extreme pH values (Tipton and Dixon, 1979).

4. The physiological pH in most animal cells is 7.0 - 7.5 at 37°C. Due to the effect of temperature, this value rises to close to 8.0 near 0°C (Scopes, 1982).

5. The buffer of choice also depends on the methods employed:
   - For gel filtration, many buffers are usually suitable.
   - For anion exchange chromatography, cationic buffers such as Tris are preferred.
   - For cation exchange or hydroxylapatite chromatography, anionic buffers such as phosphate are preferred (Blanchard, 1984).

6. Buffer mixtures with wide buffering ranges at constant ionic strength are described by Ellis and Morrison (1982).

7. A description of buffers and cryosolvents for low temperature conditions is found in Fink and Geeves (1979).

8. All chemical products should be reagent grade or higher.

B. Preparation of Buffers

- In general, the pH of the buffer should be adjusted at the temperature at which the buffer will be used. This requires that the pH electrode also be standardized at the working temperature, often 4°C. In practice, the buffer is usually prepared and the pH adjusted at room temperature (however, it should be noted that temperature effects on buffer pH may be large; note especially Tris, which at 25°C has a pK_a of 8.06, which becomes 8.85 at 0°C [Blanchard, 1984]). An experimental solution should be tested for its pH after all the components (e.g. EDTA, DTT, Mg^{2+}) have been added since the pH may be altered upon addition.

- Unless other instructions are given, assume that the pH of a buffer is adjusted down with HCl and up with either NaOH or KOH.

• The basicity of tetramethylammonium hydroxide is equivalent to NaOH or KOH. Tetramethylammonium hydroxide should be used in adjusting the pH of a buffer for a reaction which requires the complete absence of mono-, di-, or trivalent metal ions (Calbiochem "Buffers" booklet).

• If both protonated and unprotonated forms of a buffer are readily available, solutions of the two forms at the same concentration can be mixed until the desired pH is obtained, either by monitoring with a pH meter or on the basis of established tables or calculations (although verification with a pH meter is always advisable), as in the following example:

Preparation of a phosphate buffer between pH 5.8 and 7.8 (Calbiochem "Buffers" booklet):

Stock Solution A (0.2M NaH_2PO_4):
Dissolve 27.6g NaH_2PO_4 to make 1 liter in deionized water.

Stock Solution B (0.2M Na_2HPO_4):
Dissolve 28.4g Na_2HPO_4 to make 1 liter in deionized water.

pH	% A	% B
5.8	92.0	8.0
5.9	90.0	10.0
6.0	**87.7**	**12.3**
6.1	85.0	15.0
6.2	81.5	19.5
6.3	77.5	22.5
6.4	73.5	26.5
6.5	**68.5**	**31.5**
6.6	62.5	37.5
6.7	56.5	43.5
6.8	51.0	49.0

pH	% A	% B
6.9	45.0	55.0
7.0	**39.0**	**61.0**
7.1	33.0	67.0
7.2	28.0	72.0
7.3	23.0	77.0
7.4	19.0	81.0
7.5	**16.0**	**84.0**
7.6	13.0	87.0
7.7	10.5	89.5
7.8	8.5	91.5

Note: pH values are only indicative; for any set of conditions, the pH should be checked with a pH meter.

Table 1.1
pK_a Values of Common Biological Buffers

Trivial Name	Buffer Name	pK_a*
Phosphate (pK_{a1})		2.15
Citrate (pK_{a1})		3.06
Formate		3.75
Succinate (pK_{a1})		4.21
Citrate (pK_{a2})		4.76
Acetate		4.76
Pyridine		5.23
Citrate (pK_{a3})		5.40
Succinate (pK_{a2})		5.64
MES	2-(N-Morpholino)ethanesulfonic acid	6.15
Cacodylate	Dimethylarsinic acid	6.27
Carbonate (pK_{a1})		6.35
BIS-Tris	[Bis-(2-hydroxyethyl)imino]tris (hydroxymethyl)methane	6.46
ADA	N-2-Acetamidoiminodiacetic acid	6.59
PIPES	Piperazine-N,N'-bis(2-ethanesulfonic acid)	6.76
BIS-Tris propane	1,3-Bis[tris(hydroxymethyl)methylamino] propane	6.80
ACES	N-2-Acetamido-2-aminoethanesulfonic acid	6.90
Imidazole		6.95
MOPS	3-(N-Morpholino)propanesulfonic acid	7.20
Phosphate (pK_{a2})		7.20
TES	2-[Tris(hydroxymethyl)methylamino] ethanesulfonic acid	7.50
HEPES	N-2-Hydroxyethylpiperazine-N'-2-ethanesulfonic acid	7.55
HEPPS (EPPS)	N-2-Hydroxyethylpiperazine-N'-3-propanesulfonic acid	8.00
Tris	Tris(hydroxymethyl)aminomethane	8.06
Tricine	N-[Tris(hydroxymethyl)methyl]glycine	8.15
Glycylglycine		8.25
Bicine	N,N-Bis(2-hydroxyethyl)glycine	8.35
TAPS	3-{[Tris(hydroxymethyl)methyl]amino} propanesulfonic acid	8.40
Borate		9.23
Ammonia		9.25
CHES	Cyclohexylaminoethanesulfonic acid	9.55
Glycine		9.78
Carbonate (pK_{a2})		10.33
CAPS	3-(cyclohexylamino)propanesulfonic acid	10.40
Phosphate (pK_{a3})		12.43

* Values from Calbiochem "Buffers" booklet or Blanchard (1984).

C. Concentration Effects of Buffer on pH

- It is useful to prepare buffers as 10x or 100x stocks. This permits smaller storage volume, and it is possible to add a bactericidal agent (such as 0.02% sodium azide) which is diluted upon use (Scopes, 1982). Saturating solubilities of some buffers at 0°C (for full chemical names, see Table 1):

MES	0.65M
PIPES	2.3M
MOPS	3.0M
TES	2.6M
HEPES	2.3M
Tris	2.4M
Phosphate	2.5M (as K^+ salt)

- Note that dilution of concentrated stock buffer solution may change the pH. For example, a buffer with 0.1M NaH_2PO_4 and 0.1M Na_2HPO_4 is pH 6.7. Tenfold dilution raises the pH to 6.9 while after one hundredfold dilution it is 7.0 (Tipton and Dixon, 1979).

- The pH of Tris decreases by 0.1 unit per tenfold dilution (Calbiochem "Buffers" booklet).

D. Limitations of Certain Buffers

Buffers are often present at the highest concentration of all components in a protein solution and may have significant effects on a protein or enzyme. Buffers composed of inorganic compounds (phosphate, borate, bicarbonate) may interact with enzymes (or their substrates), affecting their activities. Most seriously, some buffers form coordination complexes with di- and trivalent metal ions resulting in proton release, lower pH, chelation of the metal, and formation of insoluble complexes. Buffers with low metal binding constants such as PIPES, TES, HEPES, and CAPS are preferred for studying enzymes with metal requirements (Blanchard, 1984).

- Phosphate:
    1. is a feeble buffer in the pH range 8-11;
    2. precipitates or binds many polyvalent cations;
    3. inhibits a large variety of enzymes, including kinases, phosphatases, dehydrogenases, and other enzymes with phosphate esters as substrates (Blanchard, 1984);
    4. exhibits a dependence of pK on buffer dilution (see Section C).

- Citrate binds to some proteins and forms metal complexes (Scopes, 1982).

- Cacodylate is toxic (Scopes, 1982).

- Carbonate has limited solubility and, since it is in equilibrium with CO_2, studies must be carried out in a closed system (Blanchard, 1984).

- ADA absorbs light at wavelengths up to 260nm and binds metal ions (Good et al., 1966).

- MOPS interferes with the Lowry protein assay, but not with either the Bradford or Bicinchoninic Acid assays (see Chapter 3).

- HEPES:
  1. interferes with the Lowry protein assay, but not with either the Bradford or Bicinchoninic Acid assays (see Chapter 3);
  2. as for all piperazine-based Good buffers (HEPES, EPPS, PIPES; see Good et al., 1966) forms radicals under various conditions and should be avoided in systems where redox processes are being studied (Grady et al., 1988).

- Tris:
  1. is a poor buffer below pH 7.0;
  2. possesses a potentially reactive primary amine;
  3. participates in various enzymatic reactions such as that catalyzed by alkaline phosphatase;
  4. passes through biological membranes (Calbiochem "Buffers" booklet);
  5. is affected by buffer concentration and temperature (see Section C above).

- Borate forms complexes with mono- and oligosaccharides, nucleic acids and pyridine nucleotides, and glycerol (Blanchard, 1984).

E. Preventing Buffer Contamination

- Phosphate-buffered solutions are highly susceptible to microbial contamination. However, 1M phosphate stock solutions do not usually become contaminated with bacteria (Schleif and Wensink, 1981).

- Filtering the buffer through a sterile ultrafiltration device may be useful for preventing bacterial or fungal growth, especially at pH 6-8 (Blanchard, 1984).

- To prevent buffer contamination during storage, 0.02% (3mM) sodium azide is often used. Sodium azide does not interact significantly with proteins at this concentration.

- Refrigeration helps to reduce buffer contamination.

F. Water Purity

Water is the primary ingredient in almost every laboratory solution. Most contaminating substances are removed by distillation and deionization, but traces of some compounds sometimes remain and reliable measurements with protein solutions may be affected. A description of various treatment systems for high-level purification of water for laboratory research can be found in Ganzi (1984).

III. Salts, Metal Ions, and Chelators

A. Ionic Strength: 0.1-0.2M KCl or NaCl simulates physiological conditions for many applications (O'Sullivan and Smithers, 1979).

B. Divalent Cations

If a complex is formed between the buffer and a divalent cation such as Ca^{2+} or Mg^{2+}, the capacity for buffering hydrogen ions is reduced. In addition, the availability of the metal ions to participate in an enzymatic reaction may be diminished. Thus, beware of buffers with affinities for metals.

1. Avoid Tris buffers when a metal cofactor is required for protein activity or stability. In 100mM Tris with 2mM Mn^{2+}, 29% of the metal is chelated (Morrison, 1979).

2. For purposes of reproducibility, if working with an ATP-binding enzyme, add Mg^{2+} in 1mM excess over ATP to ensure that essentially all ATP is present as Mg·ATP (Watts, 1973).

C. Chelators

When it is necessary to limit metal effects, specific metal ion chelators should be used. Metal ion chelators also inactivate metalloproteases.

1. To eliminate trace amounts of heavy metals in buffers, 0.1 - 5mM ethylene diamine tetraacetic acid (EDTA) is commonly used (Scopes, 1982).

2. The most commonly used chelating agents are EDTA and ethylene bis(oxyethylenenitrilo)tetraacetic acid (EGTA). While EDTA displays strong and nonspecific affinity for a variety of metals, EGTA's affinity for calcium is significantly higher than its affinity for magnesium, permitting the preferential sequestering of calcium in solutions with EGTA (Blanchard, 1984).

3. *o*-Phenanthroline chelates zinc, while *m*-phenanthroline does not (Todhunter, 1979).

IV. Reducing Agents

A. General Considerations

Within the cell, various reducing compounds, notably glutathione, prevent protein oxidation. Once the cell has been disrupted, care must be taken to counteract effects due to increased contact with oxygen and dilution of naturally occurring reducing agents. Many proteins lose activity when oxidized, although this activity may sometimes be restored by reduction of critical thiol groups. The presence of divalent cations may accelerate the formation of disulfide bonds (Scopes, 1982).

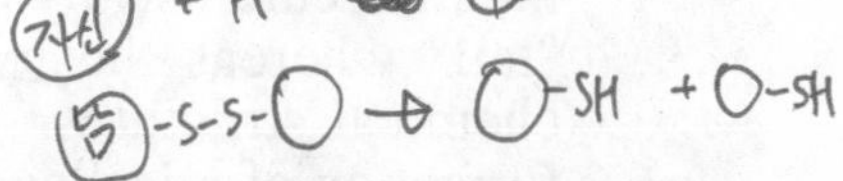

B. Specific Recommendations

- 2-Mercaptoethanol, which is easy to use since it may be stored as a solution at 4°C, must be used at a concentration of 5 - 20mM. Within 24 hours of its introduction into the buffer, 2-mercaptoethanol becomes oxidized, after which it may accelerate protein inactivation (Scopes, 1982).

- Dithiothreitol (DTT or Cleland's reagent) is supplied as a powder and must be stored at -20°C as a stock solution. DTT may be used at 0.5 - 1.0mM, and oxidation results in the formation of a stable intramolecular disulfide which does not endanger protein sulfhydryls.

- A good strategy is to use 2-mercaptoethanol at a 1:1000 dilution (about 12mM) during a protein preparation, but 1 - 5mM DTT for long term storage (Scopes, 1982).

- Note that certain enzymes are sensitive to reduction (Schleif and Wensinck, 1981).

V. Detergents

A. Introduction

Detergents are used most often for the extraction and purification of membrane proteins. Membrane proteins usually require the presence of some detergent in order to be solubilized. A number of classes of detergents may be used and some general guidelines for the solubilization and stabilization of membrane proteins are presented in this section.

Detergents are amphiphilic molecules with substantial solubility in water. With the exception of bile salts, the hydrophobic portion of the molecule usually consists of a linear or branched hydrocarbon "tail" whereas the hydrophilic "head" may have very different chemical structures. An important property of detergents is the formation of micelles, which are clusters of detergent molecules in which the hydrophilic head portions face outward. Solubilized membrane proteins form mixed micelles with detergent. The hydrophobic (or transmembrane) domain of the protein is shielded from contact with the aqueous buffer by detergent molecules. The critical micelle concentration (CMC) is the lowest detergent concentration at which micelles form. A detergent with a high CMC (i.e. octylglucoside) will return to the monomeric state upon dilution of the detergent, thus permitting rapid removal of the detergent by dialysis. In addition, the micelle molecular weight may be important for dialysis, gel filtration chromatography, and electrophoresis under non-denaturing conditions. Factors affecting the CMC include temperature, pH, ionic strength, presence of multivalent ions or organic solvents, and detergent purity.

Although the presence of high concentrations of detergent often results in protein denaturation, sometimes the subsequent removal of detergent allows the protein to renature. If protein denaturation is a problem, nonionic detergent concentrations below 0.1% are not usually harmful to proteins (Scopes, 1982).

B. Classes of Detergents

Consult Table 1.2 for specific properties of detergents.

• Ionic Detergents

Ionic detergents contain charged head groups, either positively charged (cationic detergents) or negatively charged (anionic detergents). This class of detergents has the disadvantage of being highly denaturing. However, these detergents permit the separation of proteins into their monomeric forms, thus facilitating molecular weight determination.

1. Sodium or Lithium Dodecyl Sulfate (SDS, LiDS)

For these two anionic detergents, lithium dodecyl sulfate (LiDS) has the advantage that it is soluble at 4°C while sodium dodecyl sulfate (SDS) is not. As much as 10mg LiDS or more per mg membrane protein may be necessary for complete solubilization (Boehringer Mannheim Catalog). Note that the use of these detergents in the presence of potassium buffers or ammonium sulfate may lead to their precipitation at room temperature. The critical micelle concentration is dramatically affected by the salt concentration; for SDS, the CMC drops from 8mM in the absence of salt to 0.5mM in 0.5M NaCl (Helenius et al., 1979).

2. Sodium Cholate and Sodium Deoxycholate

Sodium cholate and sodium deoxycholate (DOC) are anionic detergents that are less denaturing than other ionic detergents (Harlow and Lane, 1988). Two classes of micelles, termed primary (containing up to 9 molecules) and secondary (9 - 60 molecules) micelles, occur above the critical micelle concentration. In contrast to other detergents, free cholate or deoxycholate monomers continue to accumulate above the CMC. The pK_a of these detergents is between 8 and 9, and precipitation of the detergent in its acid form is a problem below pH 7.5. Divalent cations cause DOC precipitation.

• Nonionic Detergents

Nonionic detergents have uncharged hydrophilic head groups. As a result, they are less likely to disrupt protein-protein interactions and are particularly useful for isolating functional protein complexes. Nonionic detergents are far less denaturing than ionic detergents; however, protein aggregation may occur in the presence of these detergents.

1. Triton X-100 (Polyoxyethylene [9-10] *p-t*-octyl phenol)

Many proteins retain their activity in 1 - 3% Triton X-100. A tenfold or greater excess of Triton X-100 to membrane lipid (w/w) may be required to solubilize membrane proteins. Triton X-100 has a strong absorbance at 280nm.

2. Triton X-114 (Polyoxyethylene [7-8] *p-t*-octyl phenol)

Triton X-114 (2%) added to a protein solution has the property of causing a separation between the detergent and the aqueous phases at temperatures above 20°C, its cloud point. Hydrophilic proteins remain in the aqueous phase while integral membrane proteins may be recovered in the detergent phase (Bordier, 1981).

3. Nonidet P-40 (Polyoxyethylene [9] *p-t*-octyl phenol)

Nonidet P-40 (NP-40) absorbs relatively weakly at 280nm, and thus may be convenient for protein isolation.

4. Octylglucoside (1-O-n-octyl-β-D-glucopyranoside)

Octylglucoside (OG) is better defined chemically than Triton and therefore results may be more reproducible with this detergent. 20-45mM OG is often sufficient to solubilize membrane proteins. OG has a high CMC (see Table 1.2) and is more easily removed from a solution than is Triton.

5. Tween 20 (PEG [20] sorbitan monolaurate)

Tween 20 is commonly used to block nonspecific protein interactions in solid phase immunochemistry (ELISA, RIA, immunoblotting). Tween 20 has a very low CMC.

• Zwitterionic Detergents

Zwitterionic detergents contain head groups that possess both positive and negative charges. This class of detergents is more efficient than nonionic detergents at overcoming protein-protein interactions while causing less protein denaturation than ionic detergents.

1. CHAPS (3-[(Cholamidopropyl)dimethyl-ammonio]-1-propanesulfonate)

CHAPS does not interfere with ion exchange chromatography or isoelectric focusing. Proteins in solutions containing CHAPS may be frozen safely.

2. Zwittergent 3-14

Table 1.2

Characteristics of Common Detergents

Detergent	Molecular Weight	Critical Micelle Concentration	Concentration for Solubilization	Micelle Size
Sodium Dodecyl Sulfate	288.5[d]	7-10mM, 0.23%[d]		18 kd
Lithium Dodecyl Sulfate	272.4	6-8mM, 0.2%[a]	>10mg/mg protein[a]	
Sodium Cholate	431[f]	3-10mM, 0.2%[f]		1.8 kd
Sodium Deoxycholate	433[f]	1-2mM, 0.57%[f]		4.2 kd
Triton X-100	about 628[a]	0.3mM, 0.02%[c]	0.2-0.6mg/mg protein[b]	90 kd
Triton X-114	about 543[d]	0.35mM, 0.02%	see text (section V.B.)	
Nonidet P-40	about 603[d]	50-300μM, 0.003-0.02%[d]		90 kd

Detergent	Molecular Weight	Critical Micelle Concentration	Concentration for Solubilization	Micelle Size
Octylglucoside	292.4[a]	15-25mM,0.5%[a]	20-45mM[a]	8.0 kd
Tween 20	1230[f]	60μM, 0.006%[f]		
Tween 80	1310[f]	10μM, 0.0013%[f]		76 kd
Brij 35	1200[f]	90μM, 0.01%[e]		49 kd
CHAPS	614.9[d]	4-8mM, 0.5%[f]	6-10mM[a]	6.0 kd
Zwittergent 3-14	364[d]	0.3mM, 0.011%		30 kd

a - Boehringer Mannheim Catalog
b - Helenius and Simons, 1975
c - Hjelmeland and Chrambach, 1984
d - Calbiochem Catalog
e - Helenius et al., 1979
f - Harlow and Lane, 1988

C. Protocol for Initial Attempts at Protein Solubilization

• Experimental Steps (from Hjelmeland and Chrambach, 1984):

1. Prepare crude membrane fraction at a protein concentration of about 10mg/ml in 50mM buffer, 0.15M KCl at 4°C.

2. Prepare detergent stock solution (10%, w/v) in the same buffer.

3. Make dilutions of the crude membrane fraction (about 5mg/ml) in buffer containing the following amounts of detergent: 0.01%, 0.03%, 0.1%, 0.3%, 1%, 3%.

4. Stir gently for 1hr at 4°C (avoid foaming or sonication).

5. Centrifuge at 100,000 x g for 1hr at 4°C.

6. Remove supernatant and resuspend pellet in an equal volume of buffer containing the same detergent concentration.

7. Determine the protein concentration of each fraction (see Chapter 3).

8. Determine the enzyme activity in each fraction.

• General Comments

1. The detergent concentration which yields the highest soluble protein content and activity should be used for further studies. If unsatisfactory results are obtained with a variety of different detergents, mixtures of detergents may be tried. Low yields of protein activity may be improved by the addition of glycerol (25 - 50%, v/v), reducing agents (1mM DTT or 5mM β-mercaptoethanol), chelating agents (1mM EDTA), or protease inhibitors (75μg/ml PMSF, 20μg/ml leupeptin or pepstatin; see Section VII below).

2. Sometimes, solubilization with phosphate (0.1 - 0.2M) is successful if KCl solubilization does not work.

3. Protein solubilization very often occurs near the detergent CMC (Hjelmeland and Chrambach, 1984).

4. For separating soluble from integral membrane proteins, the method of Bordier (1981) using the nonionic detergent Triton X-114 may be employed. Extraction and separation in Triton X-114 may be used as a first step in the purification of a membrane protein (this step rapidly separates the membrane proteins from lysosomal proteases). Detergent exchange may be performed during ion-exchange chromatography.

5. The interactions of membrane-associated proteins with membranes may also be disrupted by exposing the membrane preparation to conditions of high salt (0.15 - 2M KCl), high or low pH, high doses of chelating agents (10mM EDTA or EGTA), or denaturing agents such as urea or guanidine HCl (6-10M) (van Renswoude and Kempf, 1984).

6. The definition of a soluble protein is not always clear. The ability of a protein to remain in the supernatant after a one hour centrifugation at 105,000 x g is affected by both the solution density and temperature. Another test that may be applied is gel filtration chromatography (Hjelmeland and Chrambach, 1984).

VI. Protein Environment

A. Surface Effects

Dilute protein solutions often lose activity quickly, possibly via denaturation on surfaces such as glassware, but this effect can be prevented by inclusion of high levels of another protein, commonly bovine serum albumin (BSA). Ideally, to avoid introducing a "contaminating" protein, dilute protein solutions should be rapidly concentrated. However, since enzyme reactions are sometimes assayed with protein concentrations as low as 1μg/ml which may lead to rapid inactivation, addition of BSA may be necessary. In addition, loss of the purified protein due to nonspecific adhesion onto glass surfaces (1μg of protein is absorbed on $5cm^2$ of a glass surface) is significantly diminished when the solution is supplemented with BSA. At least 0.1mg/ml BSA should be used in assay mixtures, while stored protein may contain as much as 10mg/ml BSA (Scopes, 1982).

B. Temperature

An enzyme's reaction velocity roughly doubles with a temperature increase of 10°C (for example between 18°C and 28°C) although some "cold-labile" proteins are effectively inactivated at low temperature (i.e., mitochondrial ATPase). Above 30 - 40°C, proteins vary widely in their stability, most becoming inactivated, but some remaining stable even upon boiling (i.e., bacterial alkaline phosphatase).

C. Storage

As a rule, a protein's half-life is extended by storage at low temperatures. Whether the best storage conditions are at 4°C, -20°C, -80°C, or in liquid nitrogen (-200°C), depends on the protein and its intended use. For short-term storage, for example, it may be better to store in the refrigerator than to subject the protein to freezing and subsequent thawing, which may be harmful to the protein.

Addition of glycerol has been found to be useful in stabilizing a protein during storage. 50% glycerol is often recommended for storage, while working buffers may contain 20 - 30% glycerol without creating problems for handling (Scopes, 1982). Storage in glycerol allows for maintenance of the protein solution at very low temperatures without freezing.

Freezing a protein solution is harmful in part because buffers may precipitate or be locally concentrated during freezing, leading to pH fluctuations which can be drastic (+/- 3 pH units). A high protein concentration (2mg/ml) provides some auto-buffering, and freezing a protein solution in liquid nitrogen can reduce the time needed for freezing. In general, freezing proteins in liquid nitrogen and storing at -80°C is preferable to storage at -20°C (Scopes, 1982).

Proteins are also often maintained in stable form as an ammonium sulfate precipitate or as a lyophilized powder (see Chapter 4). It is better to store a centrifuged ammonium sulfate pellet overnight than to store it in a redissolved solution.

VII. Protease Inhibitors

Disruption of cells for isolation of proteins may also result in the release of proteases from subcellular compartments. These proteases often need to be removed rapidly to ensure that the proteins remain intact. Until the protein of interest is purified, protease inhibitors should be used to retard proteolysis. Five commonly used protease inhibitors are listed below, along with indications for their use. Since protease sensitivity varies widely among different proteins, the indicated working concentration may need to be modified. Special care must be taken when diluting the protease inhibitor into the experiment buffer to mix the solution thoroughly in order to minimize precipitation of protease inhibitors with poor solubilities in aqueous solutions. A more complete list of proteases and protease inhibitors can be found in the Boehringer Mannheim catalog.

A. Common Inhibitors

- PMSF (phenylmethanesulfonyl fluoride):
  1. inhibits serine proteases (e.g., chymotrypsin, trypsin, thrombin) and thiol proteases (e.g., papain);
  2. soluble in isopropanol to 10 mg/ml;
  3. stock solution stable over a year at room temperature;
  4. working concentration: 17 - 174 µg/ml (0.1 - 1.0mM);
  5. unstable in aqueous solution, add fresh PMSF at every isolation or purification step.

- EDTA (ethylenediamine tetraacetic acid):
  1. inhibits metalloproteases;
  2. soluble in water to 0.5M at pH 8 - 9;
  3. stock solution stable for over 6 months at 4°C;
  4. working concentration: 0.5 - 1.5mM (0.2 - 0.5 mg/ml);
  5. add NaOH to adjust pH of stock solution, otherwise EDTA remains insoluble.

- Pepstatin A:
  1. inhibits acid proteases such as pepsin, renin, cathepsin D, and chymosin;
  2. soluble in methanol to 1 mg/ml;
  3. stock solution stable for 1 week at 4°C or 6 months at -20°C;
  4. working concentration: 0.7 μg/ml (1μM);
  5. insoluble in water.

- Leupeptin:
  1. inhibits serine and thiol proteases such as papain, plasmin, and cathepsin B;
  2. soluble in water to 10 mg/ml;
  3. stock solution stable for 1 week at 4°C or 6 months at -20°C;
  4. working concentration: 0.5 μg/ml (1μM).

- Aprotinin:
  1. inhibits serine proteases such as plasmin, kallikrein, trypsin, and chymotrypsin;
  2. soluble in water to 10 mg/ml, adjust to pH 7 - 8;
  3. stock solution stable for 1 week at 4°C or 6 months at -20°C;
  4. working concentration: 0.06 - 2.0 μg/ml (0.01 - 0.3μM);
  5. avoid repeated freeze-thawing;
  6. inactive at pH >12.8.

B. A Sample Broad Range Protease Inhibitor Cocktail

35μg/ml PMSF	- serine proteases
0.3mg/ml EDTA	- metalloproteases
0.7 μg/ml Pepstatin A	- acid proteases
0.5 μg/ml Leupeptin	- broad spectrum

(from Boehringer Mannheim publications)

VIII. References

Blanchard, J.S. 1984. Meth. Enzymol. 104: 404-414. Buffers for Enzymes.
Boehringer Mannheim Catalog: Biochemica Information. 1987. J. Keesey, ed. Boehringer Mannheim Biochemicals, Indianapolis.
Bordier, C. 1981. J. Biol. Chem. 256: 1604-1607. Phase Separation of Integral Membrane Proteins in Triton X-114 Solution.
Calbiochem. Buffers: A Guide for the Preparation and Use of Buffers in Biological Systems. D.E. Gueffroy, ed. 1975.
Calbiochem Biochemical/Immunochemical Catalog. 1989.
Ellis, K.J. and J.F. Morrison. 1982. Meth. Enzymol. 87: 405-426. Buffers of Constant Ionic Strength for Studying pH-Dependent Processes..
Fink, A.L. and M.A. Geeves. 1979. Meth. Enzymol. 63: 336-370. Cryoenzymology: The Study of Enzyme Catalysis at Subzero Temperatures.
Ganzi, G.C. 1984. Meth. Enzymol. 104: 391-403. Preparation of High-Purity Laboratory Water.
Good, N.E., G.D. Winget, W. Winter, T.N. Connolly, S. Izawa, and R.M.M. Singh. 1966. Biochem. 5: 467-477. Hydrogen Ion Buffers for Biological Research.
Grady, J.K., N.D. Chasteen, and D.C. Harris. 1988. Anal. Biochem. 173: 111-115. Radicals from "Good's" Buffers.
Harlow, E. and D. Lane. 1988. Antibodies: A Laboratory Manual. 726 pages. Cold Spring Harbor Laboratory, Cold Spring Harbor, New York.
Helenius, A. and K. Simons. 1975. Biochim. Biophys. Acta 415: 29-79. Solubilization of Membranes by Detergents.
Helenius, A., D.R. McCaslin, E. Fries, and C. Tanford. 1979. Meth. Enzymol. 56: 734-749. Properties of Detergents.
Hjelmeland, L.M. and A. Chrambach. 1984. Meth. Enzymol. 104: 305-318. Solubilization of Functional Membrane Proteins.
Morrison, J.F. 1979. Meth. Enzymol. 63: 257-294. Approaches to Kinetic Studies on Metal-Activated Enzymes.

O'Sullivan, W.J. and G.W. Smithers. 1979. Meth. Enzymol. 63: 294-336. Stability Constants for Biologically Important Metal-Ligand Complexes.

van Renswoude, J. and C. Kempf. 1984. Meth. Enzymol. 104: 329-339. Purification of Integral Membrane Proteins.

Schleif, R.F. and P.C. Wensink. 1981. pp. 77-78, 177-178. Practical Methods in Molecular Biology. 220 pages. New York, Springer-Verlag.

Scopes, R.K. 1982. pp. 185-193. Protein Purification: Principles and Practice. 282 pages. Springer-Verlag, New York.

Stryer, L. 1988. Biochemistry. 1089 pages. Third Edition, W.H. Freeman and Company, New York.

Tipton, K.F. and H.B.F. Dixon. 1979. Meth. Enzymol. 63: 183-234 (1979). Effects of pH on Enzymes.

Todhunter, J.A. 1979. Meth. Enzymol. 63: 383-411. Reversible Enzyme Inhibition.

Watts, D.C. 1973. Creatine Kinase. in The Enzymes P.D. Boyer, ed. Vol 8, pp.383-455. Academic Press, New York.

Chapter 2

Protein Extraction

I. Introduction

II. Homogenization
- A. General Considerations
- B. Specific Steps

III. Sonication
- A. General Considerations
- B. Specific Steps

IV. French Pressure Cell
- A. General Considerations
- B. Specific Steps

V. Grinding With Alumina or Sand
- A. General Considerations
- B. Specific Steps

VI. Glass Bead Vortexing
- A. General Considerations
- B. Specific Steps

VII. Enzymatic Treatment s
- A. General Considerations
- B. Specific Steps

VIII. Other Lysis Methods
- A. Detergent Lysis
- B. Organic Solvent Lysis
- C. Osmotic Shock Lysis
- D. Freeze-Thaw Lysis

IX. Suppliers

X. References

I. Introduction

To purify or characterize an intracellular protein, an efficient method of cell disruption must be developed which is not harmful to the protein of interest. Standard procedures for lysis of certain types of tissues or cells are available, but in some cases it is necessary to explore alternative lysis methods in order to improve protein yields or to maximize recovery of active protein. In this chapter, the principal techniques of cell lysis are described.

The first steps of a typical protein isolation procedure usually consist of washing the tissue and applying the lysis method. Then a centrifugation step separates the soluble proteins from the membrane fraction and insoluble debris. Finally, the protein sample may be analyzed, further purified, or stored.

Tissue is washed with a buffered saline solution to remove traces of blood or other extracellular material, whereas microbial cells may be centrifuged, resuspended in buffer, and recentrifuged to eliminate undesired traces of the growth media. The extract obtained after lysis, termed the **homogenate**, is usually centrifuged under conditions ranging from 10 minutes at 15,000 x g to one hour at 100,000 x g. The subsequent supernatant is called the **crude extract** and the pellet contains the membrane fraction. If the crude extract contains floating particles, the extract may be filtered through cheesecloth or glass wool. Storage of protein samples is described in Chapter 1.

A list of lysis methods for various tissues or cells is presented in Table 2.1.

Table 2.1

Cell Disruption Methods for Various Tissues

Chapter Section	Cell Lysis Method	Kind of Tissue
I.	Blade Homogenization	most animal, plant tissues
I.	Hand Homogenization (Dounce Homogenization)	soft animal tissues
II.	Sonication	cell suspensions
III.	French Pressure Cell	bacteria, yeast, plant cells
IV.	Grinding	bacteria, plant tissues
V.	Glass Bead Vortexing	cell suspensions
VI.	Enzyme Digestion	bacteria, yeast
VII.	Detergent Lysis	tissue culture cells
VII.	Organic Solvent Lysis	bacteria, yeast
VII.	Osmotic Shock	erythrocytes, bacteria
VII.	Freeze-Thaw Lysis	

II. Homogenization

A. General Considerations

One of the most common ways of disrupting soft tissues is by homogenization. Homogenization is accomplished either by chopping the tissue in a blender (Lu and Levin, 1978; Necessary et al., 1985) or by forcing the tissue through a narrow opening between a Teflon pestle and a glass container (Fleischer et al., 1979). These methods are rapid and pose relatively little danger to proteins apart from proteases liberated from other cellular compartments.

Time for lysis (from start until centrifugation): 5 - 10 min.

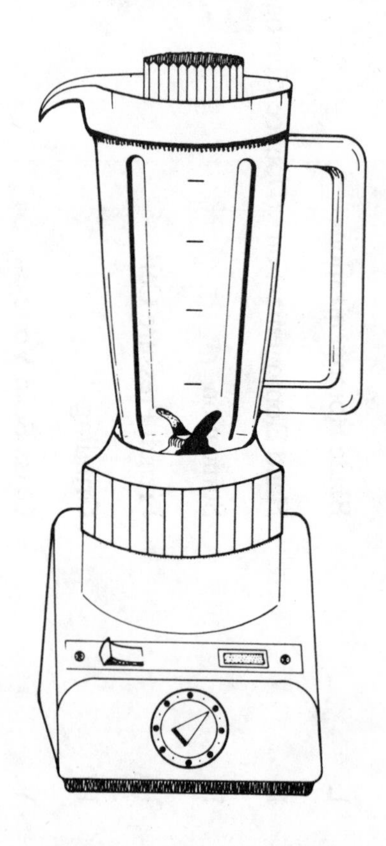

Figure 2.1. Homogenizer.

B. Specific Steps

• Protocol

1. Chop washed tissue into small pieces (for example, 1cm cubes) with a knife.

2. Add homogenization buffer, usually 3 - 5 volumes buffer per volume of tissue, and transfer to glass mortar or blender.

3. Preparation of homogenate:

    a. using a power-driven Potter-Elvehjem glass-Teflon homogenizer at 500 - 1500 rpm, pass through the sample 3 - 6 times, allowing 5 - 10 seconds per stroke.

    b. using a Dounce hand homogenizer, pass through the sample 10 - 20 times.

    c. using a blender, mix three times for 20 seconds at high speed, pausing for several seconds between each pulse.

• Comments

1. Mammalian tissues such as liver, heart, brain, and smooth muscle are commonly disrupted with the Potter-Elvehjem homogenizer. The Dounce hand homogenizer has been used for cultured cells and brain tissue.

2. The clearance between the pestle and the glass container may be varied to provide more or less shearing of tissue. Clearances may range from 0.35 to 0.70mm.

3. The glass container can be immersed in ice water to maintain a low temperature during homogenization. The blender should be prechilled to 4°C, and blending can take place in a cold room.

4. The extent of cell breakage can be monitored by observing the cells in a phase contrast microscope or by quantifying the amount of protein liberated per gram of tissue wet weight.

5. If this method does not yield satisfactory lysis, a somewhat harsher technique such as the French press or vortexing with glass beads should be tried.

III. Sonication

A. General Considerations

A sonicator disrupts tissue by creating vibrations which cause mechanical shearing of the cell wall. In order to have maximal shearing, it is necessary to tune the sonicator to achieve maximal agitation. Maximal agitation must be tempered by the need to keep the vibrations below the level where foaming of the solution occurs, since this will aerate the solution and cause protein denaturation.

The sonicator probe must be maintained below the solution surface at all times while it is turned on. To tune the sonicator prior to use, place the sonicator probe in an expendable portion of the sample (or a solution with similar viscosity) in the same size vessel as will be used for sonicating the sample. Most commonly, heavy plastic test tubes are used to provide the greatest probe-to-solution surface area. Then, the power is turned on and adjusted to a level slightly below that at which foaming occurs. At this point, the sonicator is tuned and only minor adjustments will be necessary during sonication of the sample.

Time for lysis: 5 - 10 min.

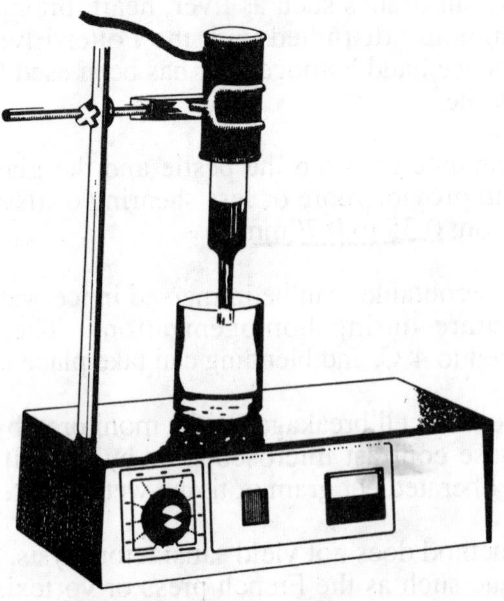

Figure 2.2. Sonicator.

B. Specific Steps

- **Protocol**

1. Suspend the washed tissue sample in at least 2 volumes of buffer. The tissue should be previously minced into small pieces with a knife or blender.

2. Sonicate for about 2 minutes. If the solution starts to foam, decrease the power setting until foaming ceases. Sonication is often carried out for several cycles (i.e., 4 x 30 sec) to permit the sample to cool on ice between treatments.

- **Comments**

1. Check for cell disruption using phase contrast microscopy after different periods of disruption by sonication.

2. Up to 1g of cells or tissue can be lysed at a time with a sonicator.

3. This method has been used for preparations from many sources, including *Escherichia coli*, *Bacillus subtilis*, *Klebsiella pneumoniae*, and homogenized brain tissue.

4. Examples for the use of sonication in membrane disruption may be found in Hochstadt (1978), Pederson and Hullihen (1979), and Enquist and Sternberg (1979).

IV. French Pressure Cell

A. General Considerations

The French Pressure Cell (French Press) achieves cell lysis by placing the sample under high pressure followed by a sudden release to atmospheric pressure. The rapid change in pressure causes cells to burst. Most often, the method is used for lysing bacteria and yeast cells. This procedure works very well for moderate volumes (10 - 30ml), but becomes technically difficult with smaller volumes and too time-consuming with larger volumes. The French pressure cell should be thoroughly cleaned before and after use to prevent contamination due to microbes, and the O-ring controlling the rate of sample release must be replaced regularly due to wear caused by the high pressure.

Time for lysis: 10 - 30 min.

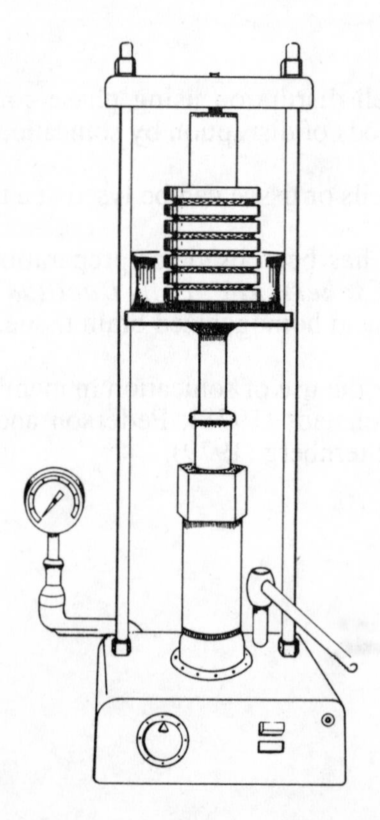

Figure 2.3. French pressure cell.

B. Specific Steps

- **Protocol**

1. Resuspend the washed cells in lysis buffer. Ratios of cell wet weight to buffer volume range from 1:1 to 1:4 g/ml.
2. Add sample to French Pressure cell and bring to desired pressure (commonly 8000 to 20,000 pounds per square inch or 550 - 1400 kg/cm^2).
3. While maintaining the pressure, adjust the outlet flow rate to 2 - 3 ml per minute (about one drop every second).
4. For some samples, a second or even third pass through the French Press is useful for greater lysis.

- **Comments**

This method has been used for preparations from many sources, including *Escherichia coli*, *Saccharomyces cerevisiae*, *Pseudomonas fluorescens*, *Paracoccus denitrificans*, and *Azotobacter vinelandii* (for examples, see Adair and Jones, 1978, and Schramm and Leung, 1978).

V. Grinding With Alumina or Sand

A. General Considerations

As with the French Press, grinding of cells with abrasive materials is most often used on unicellar organisms. The materials are inexpensive -- a mortar, pestle, and either sand or alumina -- and the procedure lends itself well to wet cell weights up to 30g. Cell lysis is achieved by the abrasive action of grinding the thick paste of sample by hand with alumina or sand (Fahnestock, 1979; Sebald et al., 1979).

Time for lysis: 5 - 15 min.

B. Specific Steps

- **Protocol**

1. To a chilled mortar, add 20g washed cell paste.
2. Prepare 40g of alumina or quartz sand to add when grinding.
3. Using the pestle, grind cell paste while gradually adding alumina or sand.
4. Grinding should continue for several minutes after the last addition of sand. It is a good sign if grinding results in snapping noises.
5. After grinding, take up cells in 60ml of buffer and centrifuge to remove cell debris and sand or alumina.

- **Comments**

This method has been used for preparations from many sources, including *Escherichia coli*, *Bacillus subtilis*, *Neurospora crassa*, *Saccharomyces cerevisiae*, as well as plant tissues.

VI. Glass Bead Vortexing

A. General Considerations

Vortexing with glass beads is in some ways an extension of the grinding method of cell lysis. The abrasive action of the vortexed beads shears cell walls, liberating the cytoplasmic contents. This method also is used primarily on unicellular organisms, particularly yeasts (Schatz, 1979). The method described is for small samples (up to 3g) and may be carried out in a test tube. For larger samples, specialized apparatus have been developed and they are listed below.

Time for lysis: 10 - 20 min.

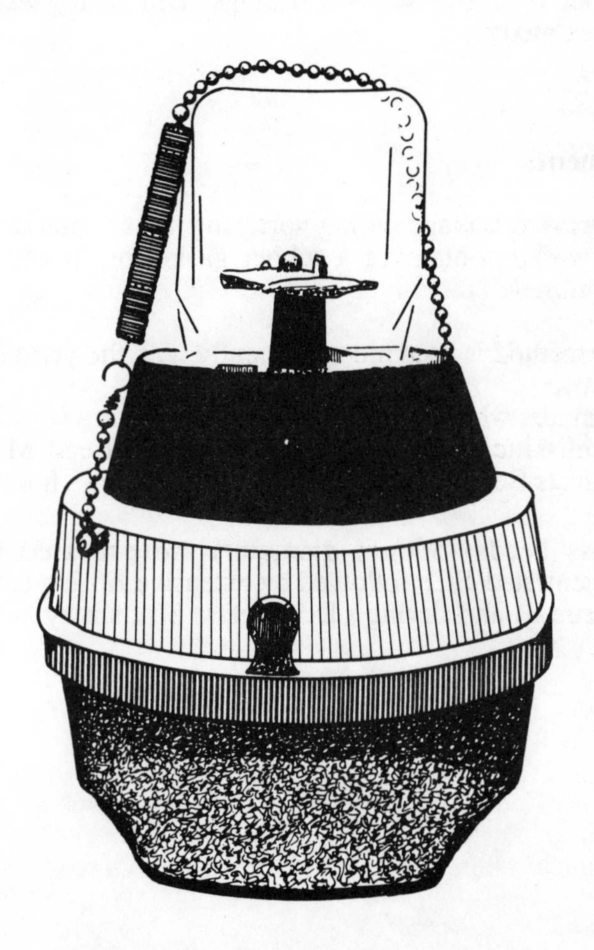

Figure 2.4. Glass-bead homogenizer.

B. Specific Steps

- **Protocol**

1. Resuspend 0.1 - 3g washed cells in an equal volume of lysis buffer and place in a sturdy tube (preferably not glass since it may shatter due to the action of the glass beads). An Eppendorf tube works well for small samples.

2. Add 1 - 3g of chilled glass beads per gram of cell wet weight.

3. Vortex 3 - 5 times for one minute, each time keeping the cells on ice for one minute between vortexings. Use the highest setting of the vortex mixer.

- **Comments**

1. To prevent leakage during vortexing, tubes should be closed with a screw cap containing a rubber gasket or should be sealed with Parafilm.

2. This method is used most frequently with the yeast *Saccharomyces*.

3. Apparatus which permit glass bead vortexing with larger quantities of cells include the Braun MSK Glass Bead Mill, the Biospec Products Bead-Beater, and the Manton-Gaulin homogenizer.

4. Glass beads (500μm diameter) are prepared by washing in concentrated HCl, followed by extensive rinsing (check that the pH is neutral) and drying. Dried glass beads may be chilled prior to use.

VII. Enzymatic Treatments

A. General Considerations

Disruption of cells by enzymatic means is principally used with microorganisms since a relatively uniform treatment is obtained when cells are in suspension. A protocol for enzymatic disruption of *Escherichia coli* follows.

Time for lysis: 15 - 30 min.

B. Specific Steps

• Protocol for *E. coli* Cells

1. Suspend washed *E. coli* cells in 3ml TE buffer per gram of cells and bring to 20 - 37°C.

   TE Buffer: 50mM Tris-Cl (8.0), 10mM EDTA

2. Add 1mg lysozyme per 5ml of cell suspension (may be added from a freshly made 10mg/ml stock solution in TE Buffer or directly as lyophilized powder into cell suspension).

3. Incubate for 10 - 20 min at 20 - 37°C, shaking gently.

• Comments

1. More rapid cell lysis may be obtained by raising the lysozyme concentration (to as much as 10mg/ml). With higher lysozyme concentrations, satisfactory lysis may be obtained in as little as 5 min even at temperatures as low as 4°C.

2. A protocol for enzymatic lysis of yeast cells can be found in Harlow and Lane (1988, pp. 455-456).

VIII. Other Lysis Methods

A. Detergent Lysis

• General Considerations

Cell lysis with detergents is commonly used with cultured animal cells (Kreibich and Sabatini, 1974; Fong et al., 1982). If low detergent concentrations are sufficient to cause cell lysis, this method may be more gentle to the protein of interest than other lysis methods.

Time for lysis: 20 - 90 min.

• Protocol

1. Wash cells several times with a buffered saline solution at 4°C (for example, TBS: 10mM Tris-Cl (pH 7.5), 150mM NaCl).
2. After the last centrifugation, resuspend the cells (10^7 - 10^8 cells/ml or 3 - 4 mg total protein/ml) in buffer containing 0.1 - 0.3% Triton X-100.
3. Vortex or invert tube several times.
4. Incubate on ice for 10 - 60 min.

• Comments

1. If the protein of interest is unstable, the incubation on ice may be eliminated. This may result in less efficient cell lysis.
2. To stabilize proteins following extraction, 0.2 volumes of 50% glycerol may be added to the extract.
3. Detergents other than Triton X-100 may also be used. Consult Chapter 1 for more information about detergents.

B. Organic Solvent Lysis

Lysis of cells using organic solvents is most commonly done with bacteria, though this method has been mainly limited to lysing cells onto filters to be incubated with antibody or nucleic acid probes (Ehrlich et al., 1979; Young and Davis, 1983).

C. Osmotic Shock Lysis

Cells are susceptible to lysis by osmotic shock when suspended in a hypotonic solution (i.e., of a lower ionic strength than the cell cytoplasm) if they are not protected by a cell wall. This method is commonly used for red blood cells.

D. Freeze-Thaw Lysis

It is possible to disrupt cells by subjecting them to several cycles of freezing and thawing. However, this method should only be tried when working with a particularly stable protein which will resist denaturation and is protected from proteolysis.

IX. Suppliers

Homogenizer: Fisher, Baxter Scientific Products
Sonicator: Fisher
French Pressure Cell: Thomas Scientific, Baxter Scientific Products
Sand: Sigma, Thomas Scientific
Glass Bead Vortexer: Biospec Products, B. Braun Instruments, Gaulin Corporation

X. References

Adair, L.B., and M.E. Jones. 1978. Meth. Enzymol. 51: 51-58. Aspartate Carbamyltransferase (*Pseudomonas fluorescens*).

Ehrlich, H.A., Cohen, S.N., and McDevitt, H.O. 1979. Meth. Enzymol. 68: 443-453. Immunological Detection and Characterization of Products Translated from Cloned DNA Fragments.

Enquist, L., and N. Sternberg. 1979. Meth. Enzymol. 68: 281-298. *In Vitro* Packaging of λ *Dam* Vectors and Their Use in Cloning DNA Fragments.

Fahnestock, S.R. 1979. Meth. Enzymol. 59: 437-443. Reconstitution of Active 50S Ribosomal Subunits from *Bacillus lichenformis* and *Bacillus subtilis*.

Fleischer, S., J.O. McIntyre, and J.C. Vital. 1979. Meth. Enzymol. 55: 32-39. Large-Scale Preparation of Rat Liver Mitochondria in High Yield.

Fong, S.-L., Tsin, A.T.C., Bridges, C.D.B., and Liou, G.I. 1982. Meth. Enzymol. 81: 133-140. Detergents for Extraction of Visual Pigments: Types, Solubilization, and Stability.

Harlow, E. and D. Lane. 1988. Antibodies: A Laboratory Manual. 726 pages. Cold Spring Harbor Laboratory, Cold Spring Harbor, New York.

Hochstadt, J. 1978. Meth. Enzymol. 51: 558-567. Adenosine Phosphoribosyltransferase from *Escherichia coli*.

Kreibich, G. and Sabatini, D.D. 1974. Meth. Enzymol. 31: 215-225. Procedure for the Selective Release of Content from Microsomal Vesicles without Membrane Disassembly.

Lu, A.Y.H, and W. Levin. 1978. Meth. Enzymol. 52: 193-200. Purification and Assay of Liver Microsomal Epoxide Hydrase.

Necessary, P.C., B. Roberts, P.A. Humphrey, G.M. Helmkamp, Jr., C.D. Turner, A.B. Rawitch, and K.E. Ebner. 1985. Anal. Biochem. 146: 372-373. The Cuisinart Food Processor Efficiently Disaggregates Tissues.

Pederson, P.L. and J. Hullihen. 1979. Meth. Enzymol. 55: 736-741. Resolution and Reconstitution of ATP Synthesis and ATP-Dependent Functions of Liver Mitochondria.

Schatz, G. 1979. Meth. Enzymol. 56: 40-50. Biogenesis of Yeast Mitochondria: Synthesis of Cytochrome c Oxidase and Cytochrome c1.

Schramm, V.L., and H.B. Leung. 1978. Meth. Enzymol. 51: 263-271. Adenine Monophosphate Nucleosidase from *Azotobacter vinelandii* and *Escherichia coli*.

Sebald, W., W. Neupert, and H. Weiss. 1979. Meth. Enzymol. 55: 144-148. Preparation of *Neurospora crassa* Mitochondria.

Young, R.A. and Davis, R.W. 1983. Proc. Natl. Acad. Sci. USA 80: 1194-1198. Efficient Isolation of Genes by Using Antibody Probes.

Chapter 3

Protein Concentration Determination

I. Absorbance at 280nm (A_{280})
- A. Summary
- B. Equipment
- C. Reagents
- D. Protocol
- E. Comments

II. Bradford Assay
- A. Summary
- B. Equipment
- C. Reagents
- D. Protocol
- E. Comments

III. Lowry Assay
- A. Summary
- B. Equipment
- C. Reagents
- D. Protocol
- E. Comments

IV. Bicinchoninic Acid (BCA) Assay
- A. Summary
- B. Equipment
- C. Reagents
- D. Protocol
- E. Comments

V. Briefly Noted: Dot Filter Binding Assay
- A. Summary
- B. Protocol

VI. Interference Table

VII. Suppliers

VIII. References

I. Absorbance at 280nm (A_{280})

A rapid method of determining whether sample solutions contain protein. Most commonly, absorbance is used for generating a protein elution profile after column chromatography.

A. Summary

- Time required: a few minutes.

- Advantages:
  1. Rapid
  2. Nondestructive

- Disadvantages:
  1. Not strictly quantitative, since this assay is based on the strong absorbance of tyrosine, phenylalanine, and tryptophan residues. Different proteins may therefore have widely varying extinction coefficients; if a protein contains no Phe, Tyr, or Trp, it will be undetected. This assay is adequate for crude protein mixtures.
  2. Strong interference by nucleic acids

- Range of sensitivity: 0.2mg/ml - 2mg/ml; can measure as little as 0.1ml (~0.05mg) in microcuvettes.

- Theory: Wetlaufer (1962)

B. Equipment

- Spectrophotometer (equipped for UV reading)
- Quartz cuvettes
- Pasteur pipets and pipet bulbs for solution transfer

C. Reagents: experiment buffer (for blank)

D. Protocol

- Single-beam spectrophotometer

  1. With experiment buffer in cuvette, set A_{280} to zero.

  2. Remove experiment buffer and add sample to cuvette, then record absorbance.

- Dual-beam spectrophotometer

  1. Set instrument zero.

  2. Add solution and buffer to sample and reference cuvettes, respectively, then record absorbance.

E. Comments

- It is a common laboratory shortcut (though a very imprecise one) to assume that an absorbance of 1.0 in a 1cm cuvette roughly approximates 1mg/ml of protein. For comparison, measured A_{280} values of a sampling of proteins at 1mg/ml follow (adapted from Whitaker and Granum, 1980):

Protein	A_{280}(1mg/ml)
Bovine Serum Albumin	0.70
Ribonuclease A	0.77
Ovalbumin	0.79
Enterotoxin	1.33
γ-Globulin	1.38
Trypsin	1.60
Chymotrypsin	2.02
α-Amylase	2.42

- This technique is most widely used for reading column fractions to obtain an estimate of where the protein peak is eluting, but unless a more specific test is performed, it is dangerous to assume that an A_{280} peak reflects eluted protein. For rapid and sensitive protein assays, try the Bradford or Dot Filter Binding Assays.

- Glass or plastic cuvettes absorb light in the UV range and should not be used for this assay.

- If absorbance is off scale, the sample can be diluted with buffer and the assay repeated. Alternatively, a cuvette with a shorter path length may be used.

- If the experiment buffer has a high absorbance relative to water, there is some interfering substance in the buffer (see Interference Table). Moderate buffer absorbance can be balanced by the zero setting, but spectrophotometers have a limited range of sensitivity at higher absorbances due to interference by stray light. See Appendix 4 for an explanation of how to test a spectrophotometer's absorbance linearity range.

- Variations of this assay which permit partial correction for interfering substances and protein composition differences include the A_{280}/A_{260} ratio (Warburg and Christian, 1941) and the A_{280}/A_{235} ratio (Whitaker and Granum, 1980).

- An approximate correction when nucleic acid is present (Schleif and Wensink, 1981):

$$\text{Protein Concentration (mg/ml)} = 1.5 \times A_{280} - 0.75 \times A_{260}$$

- A graphic representation of this method is presented in the form of a nomograph (Fig. 3.1). If the A_{280}/A_{260} is roughly 2, it can be assumed that the nucleic acid concentration is negligible.

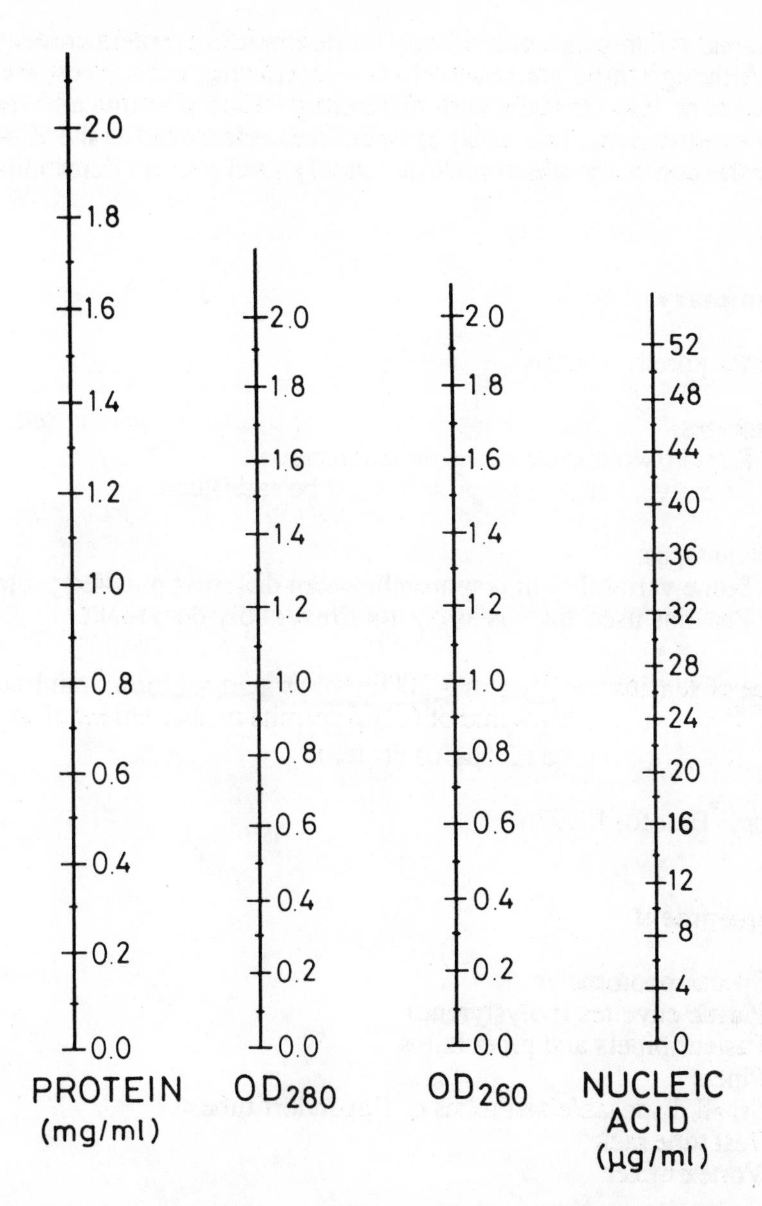

Figure 3.1. Nomograph. Alignment of a straight edge at the two points corresponding to optical density (absorbance) values on the central scales for 260nm and 280nm (obtained with a one centimeter light path cuvette) permits the estimated protein and nucleic acid concentrations to be read from the scales on the extreme left and right, respectively. Modified from an earlier version by E. Adams using the data of Warburg and Christian (1941).

II. Bradford Assay

A rapid and reliable dye-based assay for determining protein content in a solution. Although there are relatively few interfering substances, the dye interacts more or less strongly with different purified proteins and thus is not strictly quantitative. This assay is sometimes referred to as the Bio-Rad assay after the company which sells the widely used protein determination kit.

A. Summary

- Time Required: 10 min.

- Advantages:
  1. Rapid (two minute development time).
  2. Sensitive, hence little protein must be sacrificed.

- Disadvantages:
  1. Some variability in response between different purified proteins.
  2. Proteins used for this assay are irreversibly denatured.

- Range of sensitivity: 25µg/ml - 200µg/ml protein solution; minimum volume of 0.1ml permits measurement of as little as 2.5µg of protein.

- Theory: Bradford, 1976

B. Equipment

- Spectrophotometer
- Plastic cuvettes (polystyrene)
- Pasteur pipets and pipet bulbs
- Pipets
- Small disposable test tubes or Eppendorf tubes
- Test tube rack
- Vortex mixer

C. Reagents

- Serva Blue G Dye
- 1 mg/ml bovine serum albumin (BSA)
- 95% ethanol
- 85% phosphoric acid

  or Bio-Rad kit with pre-mixed reagents

D. Protocol

• Solutions

1. Bradford Stock Solution

100ml 95% ethanol
200ml 88% phosphoric acid
350mg Serva Blue G
Stable indefinitely at room temperature.

2. Bradford Working Buffer

425ml distilled water
15ml 95% ethanol
30ml 88% phosphoric acid
30ml Bradford Stock Solution
Filter through Whatman No. 1 paper, store at room temperature in brown glass bottle. Usable for several weeks, but may need to be refiltered.

Reagents are also commercially available (Bio-Rad).

• Assay

1. Pipet protein solution (maximum 100µl) into tube (see example below for standard curve).
2. Add experiment buffer to make a total volume of 100µl.
3. Add 1ml Bradford Working Buffer and vortex.
4. Read A_{595} after 2 minutes (Read and Northcote, 1981) but before 1 hour (Bearden, 1978).

- **Generating a Standard Curve**

1. A standard curve with samples of known protein concentration prepared in parallel with unknown protein solutions is essential for quantitative assessment of protein concentration. In addition, samples should be run in duplicate or triplicate. If the same cuvette is used for reading samples, it is a good idea to read those samples with less protein content first to reduce error arising from Bradford dye carryover in the cuvette as a result of incomplete rinsing.

2. Due to the continuing course of color development of the protein-dye complex, the Bradford Working Buffer should be added to standard and unknown protein solutions sequentially. All samples should also be read sequentially following the color reaction development. Each assay of unknown protein solutions must be accompanied by the generation of a new standard curve.

3. The standard curve table on the following page illustrates how to prepare a standard curve. Care must be taken to ensure that the curve remains linear over the range of protein concentrations tested or a large margin of error will be associated with unknown protein concentration readings.

4. After measuring the A_{595}, the curve presented in Fig. 3.2 is generated with the averaged values.

5. The standard curve for the Bradford Assay remains linear only from about 2.5μg to 15μg of BSA. If absorbances of unknown protein samples fall outside of this range, the margin of error becomes very high. Another consequence of the range of linearity is that a line for the standard curve should be drawn with greater emphasis placed on points in the linear range than at the edges. It is also possible to express the amount of BSA measured along the x-axis as a concentration.

Standard Curve

Sample Number	µg protein	standard solution (1mg/ml BSA)	Experiment Buffer	Bradford Reagent	A_{595}
1	0µg	0µl	100µl	1ml	0.000
2	2.5	2.5	97.5	1	0.120
3	2.5	2.5	97.5	1	0.130
4	5	5	95	1	0.250
5	5	5	95	1	0.215
6	7.5	7.5	92.5	1	0.331
7	7.5	7.5	92.5	1	0.364
8	10	10	90	1	0.460
9	10	10	90	1	0.442
10	12.5	12.5	87.5	1	0.531
11	12.5	12.5	87.5	1	0.562
12	15	15	85	1	0.633
13	15	15	85	1	0.617
14	17.5	17.5	82.5	1	0.684
15	17.5	17.5	82.5	1	0.650
16	20	20	80	1	0.721
17	20	20	80	1	0.727

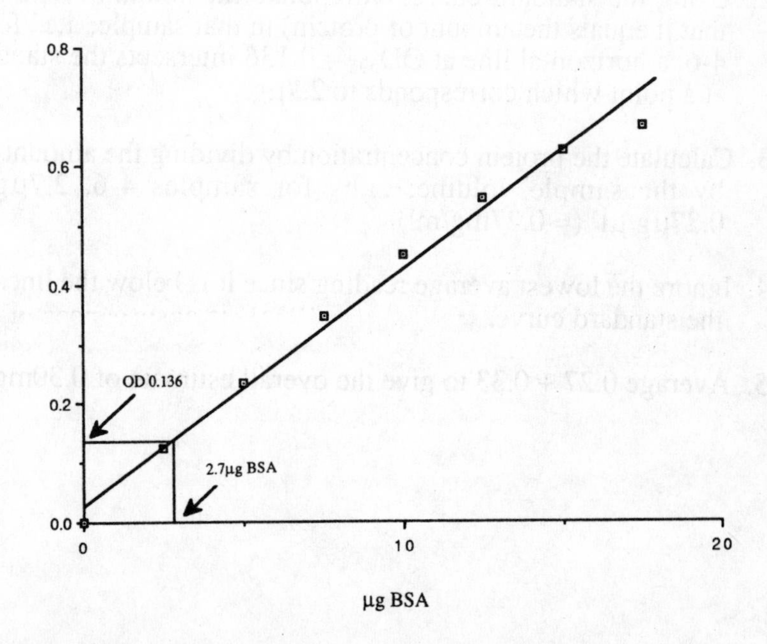

Figure 3.2. Bradford Assay Standard Curve.

- **Utilizing the Standard Curve**

Assuming an unknown protein solution yields the following readings in parallel with the standard curve above:

No.	Sample	Expt Buffer	Bradford Reagent	A_{595}	Avg A_{595}	µg protein	µg/µl (mg/ml)
1	5µl	95µl	1ml	0.059			
2	5	95	1	0.101	0.082	(unreliable)	
3	5	95	1	0.085			
4	10	90	1	0.135			
5	10	90	1	0.146	0.136	2.7	0.27
6	10	90	1	0.127			
7	50	50	1	0.700			
8	50	50	1	0.717	0.718	16.4	0.33
9	50	50	1	0.738			

The calculation of the protein concentration can be accomplished as follows:

1. Determine the average absorbance for a given volume of sample, i.e., for samples 4-6, (0.125 + 0.136 + 0.117)/3 = 0.136.

2. Using the standard curve, extrapolate the amount of BSA (assume that it equals the amount of protein) in that sample; i.e., for samples 4-6, a horizontal line at OD_{595} = 0.136 intersects the standard curve at a point which corresponds to 2.7µg.

3. Calculate the protein concentration by dividing the amount of protein by the sample volume; i.e., for samples 4-6, 2.7µg/10µl = 0.27µg/µl (= 0.27mg/ml).

4. Ignore the lowest average reading since it is below the linear range of the standard curve.

5. Average 0.27 + 0.33 to give the overall estimate of 0.30mg/ml.

E. Comments

- Color is fully developed after 5 minutes, but precipitation starts after 10-15 minutes, especially at higher protein concentrations due to the tendency for proteins to precipitate in acidic conditions. If samples are read within 10 minutes of standards, error due to color loss should be <2% (Peterson, 1983).

- Disposable plastic cuvettes are recommended because stain accumulates on cuvette walls and is difficult to remove.

- If glass cuvettes are used, clean them by rinsing thoroughly either with methanol, in concentrated glassware detergent followed by water and acetone, or by soaking in concentrated HCl overnight.

- If sample absorbance is above the useful linear range, dilute with Bradford Working Buffer to a maximum total of 5ml. The blank should be diluted with an equal volume (Spector, 1978).

- Assay results vary with different purified proteins (Pierce and Suelter, 1977, Van Kley and Hale 1977). Protein concentration measurements may be more accurate when generating the standard curve with the protein of interest or calibrating against another method (i.e., BCA, see Section IV).

- Bovine serum albumin (BSA) is most commonly used as a protein standard, although it has often been pointed out that the response of BSA to the dye is atypical. Ovalbumin is a more representative alternative standard (Read and Northcote, 1981).

- Older reagent solutions produce decreased absorbances (Spector, 1978).

- The Bradford assay can also be used for quantifying proteins immobilized on columns (Asryants et al., 1985).

- This method can be applied to membrane-bound proteins (Fanger, 1987).

III. Lowry Assay

A standard and quantitative assay for determining protein content in a solution which historically has been widely used (Lowry et al., 1951). Most of the many interfering substances can be removed by precipitating the proteins from solution prior to running the assay.

A. Summary

- Time Required: 40 min.

- Advantages:
  1. Reliable method for protein quantitation
  2. Little variation among different proteins

- Disadvantages:
  1. Many interfering substances
  2. Slow reaction rate
  3. Instability of certain reagents
  4. Proteins irreversibly denatured

- Range of sensitivity: 5μg/ml - 100μg/ml

- Theory: Peterson (1983)

B. Equipment

- Spectrophotometer
- Cuvettes
- Pipets
- Pasteur pipets and pipet bulbs
- Test tubes (3-5ml capacity)
- Test tube rack
- Vortex mixer

C. Reagents

- $CuSO_4 \cdot 5\ H_2O$
- $Na_3C_6H_5O_7$ $(\cdot 2H_20)$ (Sodium citrate)
- Na_2CO_3
- NaOH
- Folin-Ciocalteu phenol reagent

D. Protocol (from Scopes, 1982)

- Solutions

1. **Solution A**, 100ml
   0.5g $CuSO_4 \cdot 5\ H_2O$
   1g $Na_3C_6H_5O_7$ ($\cdot 2H_20$)
   Add distilled water to 100ml
   Solution may be stored indefinitely at room temperature.

2. **Solution B**, 1 liter
   20g Na_2CO_3
   4g NaOH
   Add distilled water to 1 liter
   Solution may be stored indefinitely at room temperature.

3. **Solution C**, 51ml
   1ml Solution A
   50ml Solution B

4. **Solution D**, 20ml
   10 ml Folin-Ciocalteu phenol reagent
   10ml distilled water

- **Assay**

1. Bring sample solution to 0.5ml with distilled water.
2. Add 2.5ml Solution C.
3. Vortex and let stand at room temperature for 5-10 minutes.
4. Add 0.25ml Solution D and vortex.
5. After 20-30 minutes, read A_{750}.

A sample standard curve is presented in Fig. 3.3.

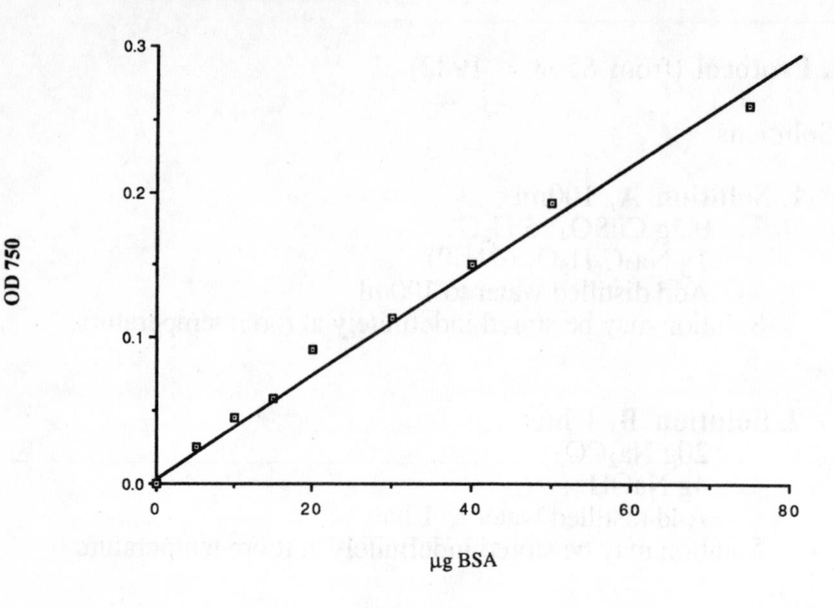

Figure 3.3. Lowry Assay Standard Curve.

- **Additional steps to purify protein sample from interfering substances:**

Deoxycholate-Trichloroacetic Acid (DOC-TCA) Precipitation

also allows concentration determination of proteins in dilute solutions (<1µg/ml) (from Peterson, 1983).

Additional Reagents:

0.15% (w/v) deoxycholate (DOC)
72% trichloroacetic acid (TCA)

1. To 1.0ml protein sample, add 0.1ml 0.15% DOC.

2. Vortex and let stand at room temperature for 10 minutes.

3. Add 0.1ml 72% TCA, vortex and spin 5-30 minutes at 1000 - 3000 x g. With fixed angle rotors, cold temperatures or large volumes, longer times are necessary.

4. Decant immediately and remove residual liquid with a pipet.

5. Redissolve pellet directly in Solution C.

E. Comments (from Peterson, 1977, 1979, 1983)

- Color development reaches a maximum in 20-30 minutes, after which there is a gradual loss of signal at about 1% per hour.

- Most interfering substances cause lower color yield, while some detergents cause a slight increase.

- High salt concentrations may cause precipitation.

- Lipids can be removed by chloroform extraction, and centrifugation may remove turbidity due to potassium ions or Triton X-100.

- Interference due to detergents, sucrose, and EDTA may be eliminated by adding SDS to the Lowry reagents (Markwell et al., 1981).

- Different protein-dye complexes generally have extinction coefficients within a factor of 1.2 of one another.

IV. Bicinchoninic Acid (BCA) Assay

A recently developed variation of the Lowry assay. The reaction is simpler to perform and has fewer interfering substances.

A. Summary

- Time Required: 40 min, 2 hrs, or overnight.

- Advantages:
  1. Single reagent
  2. End product is stable
  3. Fewer interfering substances than Lowry assay

- Disadvantages:
  1. Slow reaction time
  2. Proteins irreversibly denatured

- Range of Sensitivity:
  Standard Assay: 10-1200 μg/ml
  Micro-assay: 0.5-10 μg/ml

- Theory: Smith et al. (1985)

B. Equipment

- Spectrophotometer
- Water bath at 37°C (optional)
- Cuvettes
- Pipets
- Pasteur pipets and pipet bulbs
- Test tubes
- Test tube racks

C. Reagents

- BCA (bicinchoninic acid)
- $Na_2CO_3 \cdot H_2O$ (sodium carbonate)
- $Na_2C_4H_4O_6$ ($\cdot 2H_20$) (sodium tartrate)
- NaOH (sodium hydroxide)
- $NaHCO_3$
- $CuSO_4 \cdot 5\ H_2O$ (copper sulfate)

D. Protocol (from Smith et al., 1985)

• Solutions

1. **Reagent A**, 1 liter
   10g BCA (1%)
   20g $Na_2CO_3 \cdot H_2O$ (2%)
   1.6g $Na_2C_4H_4O_6$ ($\cdot 2H_2O$) (0.16%)
   4g NaOH (0.4%)
   9.5g $NaHCO_3$ (0.95%)
   add distilled water to 1 liter
   If needed, add NaOH or solid $NaHCO_3$ to adjust pH to 11.25.

2. **Reagent B**, 50ml
   2g $CuSO_4 \cdot 5\ H_2O$ (4%)
   add distilled water to 50ml

Reagents A and B are stable for at least 12 months at room temperature, and are commercially available (Pierce).

3. **Standard Working Reagent (SWR)**
   50 volumes Reagent A
   1 volume Reagent B
   Stable for 1 week (Pierce Bulletin)

• **Assay**

1. Mix 1 volume of sample with 20 volumes of SWR (i.e. 100µl and 2ml).

2. Incubate (a) at room temperature for 2 hours or (b) at 37°C for 30 minutes.

3. Cool to room temperature in the case of (b).

4. Read A_{562}.

A standard curve is presented in Fig. 3.4.

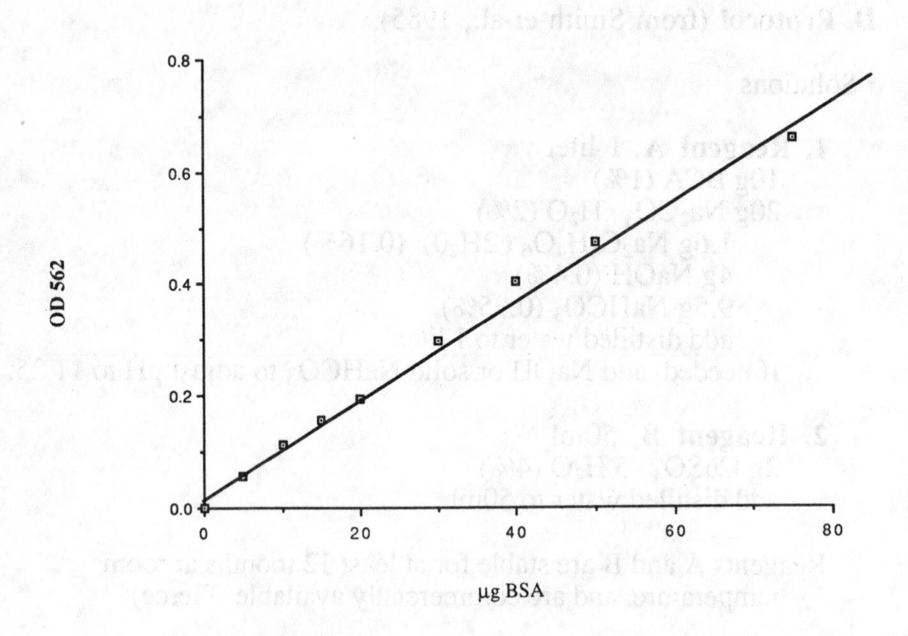

Figure 3.4. BCA Assay Standard Curve.

E. Comments

- After cooling sample to room temperature, absorbance continues to increase at about 2.3%/10min (Pierce BCA Handbook).

- The BCA micro-assay with a detection sensitivity of 0.5 - 10μg/ml utilizes more concentrated reagents and requires a 60 minute development time at 60°C (Pierce BCA Handbook).

- Samples containing lipids display aberrantly high absorbances with the BCA assay, even relative to the Lowry assay (Kessler and Fanestil, 1986).

- BCA assay variation for protein determination in buffers with sulfhydryl reagents and detergent can be found in Hill and Straka, 1988.

- Application of the deoxycholate-trichloroacetic acid (DOC-TCA) precipitation protocol (see Section III.D.) to the protein sample will remove most interfering substances.

V. Briefly Noted: Dot Filter Binding Assay

A. Summary

Dot filter binding (Coluccio and Bretscher, 1987) is a rapid way to obtain a preliminary determination of protein content in a large number of samples. This assay is especially recommended for screening fractions from density gradient centrifugation or column chromatography. The information obtained from this initial screening can be used as a guide for pooling fractions from a protein peak and will reduce the number of samples which require more quantitative protein concentration determination.

B. Protocol

- **Materials**

1. Whatman 3MM paper
2. 10% Trichloroacetic Acid (TCA)
3. Coomassie Gel Stain (see Chapter 5)
4. Coomassie Gel Destain (see Chapter 5)

- **Assay**

1. Using a pencil, prepare a 1cm x 1cm grid on 3MM paper.

2. Apply 3µl of each sample in the center of the grid square assigned to the sample.

3. Allow filter paper to dry well, about 15 minutes.

4. Soak filter paper thoroughly in 10% TCA for 30 seconds (TCA solution may be reused).

5. Rinse paper briefly with water.

6. Stain filter paper in Coomassie Gel Stain for 30 seconds (stain may be reused).

7. Rinse paper briefly with water.

8. Destain filter paper in Coomassie Gel Destain. Destaining should be complete within a few minutes.

9. Identify fractions containing significant amounts of protein by the appearance of blue color in the corresponding squares.

VI. Interference Table

Highest Acceptable Concentrations

Compound	A_{280}	Bradford	Lowry	BCA
Buffers				
Potassium phosphate	OK	1M[1]	0.03M[5]	-
Sodium phosphate	OK	1M[1]	0.25M[5]	0.1M[6]
Tris	OK	2M[1]	250µM[5]	0.1M[6]
Na^+-citrate	-	50mM[1]	2.5mM[5]	-
HEPES	-	100mM[1]	2.5µM[5]	100mM[6]
PIPES	OK	500mM[1]	5µM[5]	50mM[7]
MOPS	-	200mM[1]	25µM[5]	50mM[7]
MES	-	700mM[1]	25µM[5]	50mM[7]
TES	-	-	1mM[5]	50mM[7]
Cacodylate-Tris	-	0.1M[9]	-	-
glycine	-	0.1M[1]	2.5mM[5]	1M, pH11[6]
Salts				
urea	-	6M[1]	200mM[5]	3M[6]
NaCl	OK	1M[1]	30mM[5]	-
ammonium sulfate	-	1M[1]	28mM[5]	interferes[8]
sodium acetate	-	-	-	0.2M, pH5.5[6]
Detergents				
Sodium deoxycholate	-	0.25%[1]	0.0625%[5]	-
guanidine-HCl	-	OK[1]	-	4M[8]
Triton X-100	interferes	0.1%[1]	0.25%[5]	1%[6]
SDS	OK	0.1%[1]	1.25%[5]	1%[6]
Nonidet P-40	interferes	interferes[2]	-	1%[6]
Tween-20	-	interferes[2]	-	-
octylglucoside	-	2%[2]	-	1%[6]
Sodium cholate	-	interferes[2]	-	-
CHAPS	OK[11]	1%[1]	1mM[11]	1%[6]
Brij 35	-	interferes[2]	-	1%[6]

(*continued*)

VI. Interference Table (continued)

Highest Acceptable Concentrations

Compound	A_{280}	Bradford	Lowry	BCA
Sugars				
sucrose	OK	-	10mM[5]	1M[6]
glucose	OK	OK[1]	30mM[5]	10mM[6]
Chelators				
EGTA	OK	0.05M[1]	interferes	-
EDTA	OK	100mM[1]	125µM[5]	10mM[6]
Reducing Agents				
2-mercaptoethanol	-	1M[1]	1.75mM[5]	50µM[10]
DTT	interferes in oxidized state	1M[1]	50µM[5]	1mM[6]
Alcohols, Polar Compounds				
ethylene glycol	OK	-	0.25%[5]	-
ethanol	OK	OK[1]	12.5%[5]	-
acetone	interferes	OK[1]	1.25%[5]	-
glycerol	OK	99%[1]	25%[5]	10%[6]
DMSO	OK	-	6.2%[5]	-
methanol	OK	OK[1]	-	-

(continued)

VI. Interference Table (continued)

Highest Acceptable Concentrations

Compound	A_{280}	Bradford	Lowry	BCA
Miscellaneous				
$MgCl_2$	OK	1M[1]	-	-
lipids	-	-	25µM[5]	interferes[9]
acrylamide	-	-	1.25mg/ml[5]	-
DNA	interferes	1mg/ml[1]	190µg/ml[5]	-
RNA	interferes	0.3mg/ml[1]	-	-
ATP	interferes	1mM[1]	-	-
ampholytes	-	1%[3,4]	interferes[5]	-
TCA-neutralized	-	-	12.5mg/ml[5]	-
HCl	OK	-	-	0.1N[6]
NaOH	OK	-	-	0.1N[6]
NAD	interferes	1mM[1]	-	-
phenol	interferes	5%[1]	-	-
amino acids	aromatics interfere	OK[1]	-	-
polypeptides<3kd	interferes	OK[1]	-	-

A dash (-) indicates that this compound has not been tested.

"OK" indicates that this compound was used successfully but no concentration was indicated.

Note that there may be some variation in permissible amounts due to differences in the assay protocol.

1 - Bio-Rad Bulletin No. 1069. 1987.
2 - Fanger, 1987.
3 - Read and Northcote, 1981.
4 - Spector, 1978.
5 - Peterson, 1983.
6 - Pierce Technical Bulletin 23225, 1984.
7 - Kaushal and Barnes, 1986.
8 - Smith et al., 1985.
9 - Kessler and Fanestil, 1986.
10 - Hill and Straka, 1988.
11 - Boehringer Mannheim Catalog

VII. Suppliers

Bradford Assay
- BioRad (kit)
- Fluka
- Pierce (kit)
- Serva
- Sigma

Lowry Assay
- Fluka
- Merck
- Sigma

Bicinchoninic Acid Assay
- Fluka
- Pierce (kit)
- Sigma

Dot Filter Binding Assay
- Sigma
- Whatman

VIII. References

A_{280}

Schleif, R.F. and P.C. Wensink. 1981. Practical Methods in Molecular Biology. New York, Springer-Verlag. p74.

Warburg, O. and W. Christian. 1941. Biochem. Z. 310: 384-421. Isolierung und Kristallisation des Gaerungsferments Enolase.

Wetlaufer, D.B. 1962. Adv. Prot. Chem. 17: 303-390. Ultraviolet Spectra of Proteins and Amino Acids

Whitaker, J.R. and P.E. Granum. 1980. Anal. Biochem. 109: 156-159. An Absolute Method for Protein Determination Based on Difference in Absorbance at 235 and 280 nm.

Bradford

Asryants, R.A., I.V. Duszenkova, and N.K. Nagradova. 1985. Anal. Biochem. 151: 571-574. Determination of Sepharose-Bound Protein with Coomassie Brillient Blue G-250.

Bearden, J.C. 1978. Biochem. Biophys. Acta 533: 525-529. Quantitation of Submicrogram Quantities of Protein by an Improved Protein-Dye Binding Assay.

Bio-Rad Protein Assay. 1987. Bio-Rad Technical Bulletin 1069, Bio-Rad Laboratories.

Bradford, M.M. 1976. Anal. Biochem. 72: 248-254. A Rapid and Sensitive Method for the Quantitation of Microgram Quantities of Protein Utilizing the Principle of Protein-Dye Binding.

Fanger, B.O. 1987. Anal. Biochem. 162: 11-17. Adaptation of the Bradford Protein Binding Assay to Membrane-Bound Proteins by Solubilizing in Glucopyranoside Detergents.

Peterson, G.L. 1983. Meth. Enzymol. 91: 95-121. Determination of Total Protein.

Pierce, J. and C.H. Suelter. 1977. Anal. Biochem. 81: 478-480. An Evaluation of the Coomassie Brilliant Blue G-250 Dye-Binding Method for Quantitative Protein Determination.

Read, S.M. and D.H. Northcote. 1981. Anal. Biochem. 116: 53-64. Minimization of Variation in the Response to Different Proteins of the Coomassie Blue G Dye-Binding Assay for Protein.

Spector, T. 1978. Anal. Biochem. 86: 142-146. Refinement of the Coomassie Blue Method of Protein Quantitation.

Van Kley, H. and S.M. Hale. 1977. Anal. Bioch. 81: 485-487. Assay for Protein by Dye Binding.

Lowry

Lowry, O.H., N.J. Rosebrough, A.L. Farr, and R.J. Randall. 1951. J. Biol. Chem. 193: 265-275. Protein Measurement with the Folin Phenol Reagent.

Markwell, M.A.K., S.M. Haas, N.E. Tolbert, and L.L. Bieber. 1981. Meth. Enzymol. 72: 296-303. Protein Determination in Membrane and Lipoprotein Samples.

Peterson, G.L. 1977. Anal. Biochem. 83: 346-356. A Simplification of the Protein Assay Method of Lowry et al. Which is More Generally Applicable.

Peterson, G.L. 1979. Anal. Biochem. 100: 201-220. Review of the Folin Phenol Protein Quantitation Method of Lowry,..

Peterson, G.L. 1983. Meth. Enzymol. 91: 95-121. Determination of Total Protein.

Scopes, R.K. 1982. Protein Purification: Principles and Practice. New York, Springer-Verlag. pp. 240, 265-266.

BCA

Hill, H.D. and J.G. Straka. 1988. Anal. Biochem. 170: 203-208. Protein Determination Using Bicinchoninic Acid in the Presence of Sulfhydryl Reagents.

Kaushal, V. and L.D. Barnes. 1986. Anal. Biochem. 157: 291-294. Effect of Zwitterionic Buffers on Measurement of Small Masses of Protein with Bicinchoninic Acid.

Kessler, R.J. and D.D. Fanestil. 1986. Anal. Biochem. 159: 138-142. Interference by Lipids in the Determination of Protein Using Bicinchoninic Acid.

Pierce Chemical Company (1984) BCA Protein Assay Reagent, Technical Bulletin 23225, Rockford, Ill.

Smith, P.K, R.I. Krohn, G.T. Hermanson, A.K. Mallia, F.H. Gartner, M.D. Provenzano, E.K. Fujimoto, N.M. Goeke, B.J. Olson, and D.C. Klenk. 1985. Anal. Biochem. 150: 76-85. Measurement of Protein Using Bicinchoninic Acid.

Dot Filter Binding Assay

Coluccio, L.M. and A. Bretscher. 1987. J. Cell Biol. 105: 325-333. Calcium-regulated Cooperative Binding of the Microvillar 110K-Calmodulin Complex to F-Actin.

Chapter 4

Concentrating Protein Solutions

I. Analytical Methods

The first two methods for protein concentration, acid and organic precipitations, result in protein denaturation. The reduced solubility of denatured proteins allows for their recovery in a pellet following centrifugation.

A. Trichloroacetic Acid (TCA) Precipitation

- **Equipment**

1. Eppendorf tubes
2. Eppendorf centrifuge
3. Vortex mixer

- **Reagents**

1. Trichloroacetic Acid (TCA), 100%
2. NaOH, 0.1N (400mg NaOH in 100ml H_2O)

- **Protocol** (Hames, 1981)

1. To a 1ml sample containing at least 5µg of protein, add 100µl of 100% trichloroacetic acid (TCA) and vortex.

2. Allow protein to precipitate 30 minutes on ice (or 15 minutes in the freezer).

3. Spin for 5 minutes at 10,000 x g.

4. Remove supernatant by decanting immediately and aspirating remaining liquid with a pipet.

5. Resuspend in 50 - 100µl 0.1N NaOH and vortex.

- **Comments**

1. Minimum protein concentration required for TCA precipitation is 5µg/ml.

2. It is sometimes useful to wash the TCA pellet with 10% TCA or with ethanol-ether (1:1) to remove the TCA (Hames, 1981).

3. Instead of resuspending the TCA precipitate in NaOH (step 5), it may be more practical to carry out an ethanol-ether (1:1) wash and resuspend the pellet in a buffered solution.

4. Deoxycholate-trichloroacetic acid (DOC-TCA) precipitation is a common variation of the TCA precipitation which may extend the precipitation range of protein concentration to below 1µg/ml (Peterson, 1983).
   a. To a 1ml sample, add 0.1ml 0.15% DOC.
   b. Vortex and incubate at room temperature for 10 minutes.
   c. Add 50µl 100% TCA, vortex and go to step 3 (Peterson, 1977).

5. DOC-TCA precipitation may not work in the presence of SDS (Cabib and Polacheck, 1984).

B. Acetone Precipitation

• **Equipment**

1. Eppendorf tubes
2. Eppendorf centrifuge
3. Vortex mixer

• **Reagents:** Acetone (or other organic solvent such as ethanol or methanol)

• **Protocol** (Hames, 1981)

1. Add 1ml of cold acetone (-20°C) to 200μl of sample solution and vortex.

2. Incubate at -20°C for 10 minutes.

3. Centrifuge for 5 minutes in an Eppendorf centrifuge.

4. Remove supernatant and air dry pellet.

5. Resuspend pellet in 1-2 pellet volumes of buffer.

• **Comments**

1. Protein pellet can be washed by repeated acetone precipitation.

2. Quantitative precipitation of <1μg of protein can be achieved by extending the -20°C incubation to >2 hours and spinning for 10 minutes at 27,000 x g (Sargent, 1987).

3. Neutral salts increase protein solubility, while divalent cations reduce solubility in organic solutions (Kaufman, 1971).

C. Immunoprecipitation

A protein-specific antibody may permit quantitative isolation of the protein of interest by immunoprecipitation (Lerner and Steitz, 1979). This procedure consists of three steps. First, the specific antibody is added to the cell extract. In order to provide the mass necessary for precipitation from solution, chemically fixed *Staphylococcus aureus* bacteria are then added. These bacteria form complexes with the antibody via protein A, a bacterial protein which has a high affinity for the F_c portion of immunoglobulins. Alternatively, purified protein A coupled to beads of Sepharose offers a solid matrix for removing the antibody-antigen complex from the cell extract. Finally, a thorough wash of the pellet removes unprecipitated material.

Staphylococcus aureus cells should be washed prior to use for immunoprecipitation in order to remove damaged or poorly fixed cells (Kessler, 1981). If protein A-Sepharose beads are used, they should be swelled prior to use according to the manufacturer's instructions.

- **Equipment**
  1. Eppendorf tubes
  2. Eppendorf centrifuge
  3. Vortex mixer

- **Reagents**
  1. Protein-specific antibody
  2. Fixed *Staphylococcus aureus* cells or Protein A-Sepharose Beads
  3. Tris
  4. NaCl
  5. EDTA
  6. NaN_3
  7. Nonidet P-40

Buffer A
50mM Tris-HCl (pH 7.4)
150mM NaCl
5mM EDTA
0.02% NaN_3

- **Washing *S. aureus* cells**

1. Within 24 hours of use, centrifuge the necessary volume of cells for 3 minutes in an Eppendorf centrifuge or at 3000 x g for 15 minutes.

2. Resuspend in original volume with Buffer A containing 0.5% Nonidet P-40.

3. Centrifuge as above.

4. Resuspend in original volume with Buffer A containing 0.05% Nonidet P-40.

5. Centrifuge as above.

6. Resuspend in original volume with Buffer A.

These washes can be carried out during step 2 of the immunoprecipitation protocol (see below).

- **Immunoprecipitation Protocol** (Kessler, 1981)

1. Add a molar excess of specific antibody to the protein solution containing the antigen of interest. Typically, 10 - 25μg of purified antibody are used in a 1ml precipitation mix.

2. Vortex and incubate precipitation mix on ice for at least 1 hour (2 - 3 hours for hybridoma supernatants).

3. Add a sufficient quantity of *Staphylococcus aureus* cells or protein A-Sepharose beads to bind all immunoglobulin molecules in the precipitation mix. Calbiochem Standardized Pansorbin Cells provide an immunoglobulin binding capacity of 2.0mg per ml of cells. Bio-Rad Affi-Gel protein A binds 6 - 7mg of IgG per ml of gel.

4. Vortex and incubate on ice for 15 - 60 minutes.

5. Centrifuge immunoprecipitate in an Eppendorf centrifuge for 15 seconds or spin at 3000 x g for 10 minutes at 4°C.

6. Wash immunoprecipitate 3 - 5 times in Buffer A containing 0.05% Nonidet P-40, taking care to resuspend the pellet completely with each wash.

   a. Resuspend pellet in 100μl washing buffer by vortexing.
   b. Centrifuge for 15 seconds in an Eppendorf centrifuge.
   c. Repeat steps **a** and **b** 3 - 5 times.

7. Pellets can be resuspended in 100μl of sample buffer (see Chapter 5, section I.C.) or isoelectric focusing buffer (see Chapter 7, Section I.G.). A quarter to a half of the pellet is often sufficient for detection on a gel, although this depends on the starting protein concentration. Heating to 100°C will be necessary for efficient separation of the protein complex.

• Comments

1. The antibody solution should be ultracentrifuged (160,000 x g for 30 minutes in an airfuge) prior to use in order to remove aggregates (Kessler, 1981). Repeated freeze-thawing of antibody solutions will lead to increased aggregation and thus the antibody should be stored in aliquots. Antibody may be incubated at 56°C for 30 minutes prior to use in order to inactivate proteases (Scheidtmann, 1989).

2. *Staphylococcus aureus* cells may be cultured and prepared for immunoprecipitation (see Kessler, 1981) or purchased already formalin-fixed and heat-treated. Prepared cells may be stored as aliquots at -20°C.

3. Nonspecific *S. aureus* binding can be reduced by preabsorbing washed *S. aureus* cells with antigen extract.

4. The specificity of the immunoprecipitate may be improved by changing the pH, salt concentration or detergent utilized in the washing buffer.

5. *Staphylococcus aureus* has different binding affinities for different immunoglobulins (Schantz, 1983). Good binding is found with rabbit, human, and guinea pig immunoglobulins. Differential binding with different immunoglobulin subclasses is seen with the following animals:

	Good	Poor
Mouse	IgG_{2a}, IgG_{2b}, IgG_3	IgG_1
Rat	IgG_1, IgG_{2c}	IgG_{2a}, IgG_{2b}
Goat	IgG_2	IgG_1
Sheep	IgG_2	IgG_1

For mouse and goat antibodies, maximal binding is found above pH 8 and 9, respectively. To compensate for poor mouse immunoglobulin subclass 1 binding, for example, it is possible to preincubate the *S. aureus* cells with anti-mouse IgG antibody.

6. Protein G is isolated from a *Streptococcus* strain and is similar to protein A in its properties of binding antibodies. Protein G possesses better binding properties than protein A to antibodies from various species (Harlow and Lane, 1988). A protein G derivative with the albumin binding domain deleted has been coupled to Sepharose and is available commercially (Pierce, Pharmacia-LKB).

7. Immunoprecipitates may require urea in sample buffer to aid in solubilization.

8. Conditions used for antigen release following immunoprecipitation include 3.5M $MgCl_2$, 0.1M NaCitrate (pH 3.0), 3M ammonium or potassium thiocyanate, saturated guanidine hydrochloride, 0.2M lithium diiodosalicylate, and 2-mercaptoethanol.

9. This procedure is most often used for isolating radiolabeled proteins and is followed by autoradiography (Kessler, 1981). Alternatively, the immunoprecipitate may be detected by immunoblotting; however, controls must be included to rule out cross-reactivity with *S. aureus* proteins.

10. Suggestions for running proper controls and reducing background precipitation can be found in Harlow and Lane (1988, pp. 465, 469-470).

II. Preparative Methods

Methods for concentrating proteins which permit retention of protein activity are discussed in the following portion of this chapter.

A. Ammonium Sulfate Precipitation ("Salting Out")

When high concentrations of salt are present, proteins tend to aggregate and precipitate out of solution. This technique is referred to as "salting out". Since different proteins precipitate at different salt concentrations, salting out is often used during protein purification. It is important to remember that factors such as pH, temperature and protein purity play important roles in determining the salting out point of a particular protein (theory: Scopes, 1981).

Ammonium sulfate is the salt of choice because it combines many useful features such as salting out effectiveness, pH versatility, high solubility, low heat of solution and low price (Scopes, 1981).

Ammonium sulfate concentrations are generally expressed in percent saturation, and a simple equation for calculation of grams of ammonium sulfate needed to make an X% solution starting from X_o% is:

$$g = \frac{515\,(X - X_o)}{100 - 0.27X}$$

(for a 1 liter solution at 0°C); see also Appendix 3

Since most proteins will precipitate at 55% ammonium sulfate, a good value for obtaining maximum protein precipitation is 85%. For a 100ml solution containing no ammonium sulfate at the start, the following protocol is recommended:

• **Equipment**

1. Beakers
2. Magnetic stir plate and stir bar
3. Centrifuge and rotor

• **Reagents**

1. Ammonium sulfate
2. Buffer in which to resuspend pellet

• **Protocol**

1. Place beaker of the protein solution in a cooling bath on top of a magnetic stir plate. This can be accomplished by placing the beaker within another beaker containing a water - ice slurry.

2. While agitating gently on a magnetic stirrer, slowly add 56.8g ammonium sulfate. Add salt more slowly as final saturation is approached. This step should be completed in 5 - 10 minutes.

3. Continue stirring for 10 - 30 minutes after all salt has been added.

4. Spin at 10,000 x g for 10 minutes or at 3000 x g for 30 minutes.

5. Decant supernatant and resuspend precipitate in 1 - 2 pellet volumes of buffer. Any insoluble material remaining is probably denatured protein and should be removed by centrifugation.

6. Ammonium sulfate can be removed by dialysis, ultrafiltration, or a desalting column.

• **Comments**

1. Stirring must be regular and gentle. Stirring too rapidly will cause protein denaturation as evidenced by foaming. It is also important to use a magnetic stirrer which does not generate a significant amount of heat while stirring.

2. While most proteins precipitate in the first 20 minutes after the salt is dissolved, some precipitation continues for hours.

3. Precipitation should be carried out in a buffer of at least 50mM in order to compensate for a slight acidification upon dissolving ammonium sulfate.

4. The buffer should contain a chelating agent such as EDTA to remove possible traces of metal in the ammonium sulfate which might be detrimental to the protein of interest.

5. To ensure maximal precipitation, it is best to start with a protein concentration of at least 1mg/ml.

6. Ammonium sulfate precipitation is often a good way of stabilizing proteins for storage (Scopes, 1981).

7. Few proteins precipitate below 24% ammonium sulfate while most do by 80%. Frequently, ammonium sulfate precipitation results in removal of RNA and DNA (Schleif and Wensink, 1981).

8. A protein's solubility may be reduced at its isoelectric point where electrostatic interactions can lead to protein aggregation and precipitation (Scopes, 1981). A lower ammonium sulfate concentration will be required to precipitate a protein at its isoelectric point.

B. Organic Solvent Precipitation

Organic solvents such as acetone and ethanol have effects similar to high levels of salt when added to protein solutions; that is, they lower the protein solubility. Proteins are more easily denatured in organic solvents at temperatures above 10°C, so special care must be taken to work with chilled solutions and rotors (0 - 4°C is satisfactory). Ionic strengths between 0.05 and 0.2M are recommended (Kaufman, 1971; Scopes, 1982).

Concentrations of organic solvents are generally calculated in percent, assuming that volumes are additive which is not strictly the case. Thus, to bring a 100ml aqueous solution to 50% (v/v) ethanol, one would add 100ml of ethanol even though the final volume only measures 192ml. Most proteins larger than 15 kd precipitate with 50% organic solvent.

- **Equipment**
  1. Beaker
  2. Magnetic stir plate and stir bar
  3. Centrifuge and rotor

- **Reagents**
  1. Acetone or ethanol
  2. Buffer in which to resuspend pellet

- **Protocol**

  1. To 100ml of protein solution at 4°C, slowly add 100ml acetone or ethanol (chilled to -20°C) with continuous gentle stirring on ice.

  2. Continue stirring on ice for 10 - 15 minutes.

  3. Centrifuge in a cold rotor for 10 minutes at 10,000 x g.

  4. Decant supernatant.

  5. Gently resuspend pellet in two pellet volumes of cold buffer. Particles which are difficult to dissolve are likely to be denatured proteins and should be removed by filtration or centrifugation.

• **Comments**

1. Acidic proteins may be more easily precipitated as complexes with divalent cations such as magnesium.

2. Larger proteins as well as more hydrophilic ones tend to come out of solution at lower organic solvent concentrations.

3. A pH closer to the protein's isoelectic point reduces its solubility in organic solvents.

C. Polyethylene Glycol (PEG) Precipitation

Polyethylene glycol, a nonionic water-soluble polymer, causes little protein denaturation while inducing precipitation at a discrete PEG concentration for a given protein (Ingham, 1984). This property, along with its low heat of solution and short equilibration time for precipitation, makes it a useful reagent for protein fractionation. Maximum protein precipitation is generally achieved with a final PEG concentration of 30%.

- **Equipment**

1. Beaker
2. Magnetic stir plate and stir bar
3. Centrifuge and rotor

- **Reagents:** Polyethylene Glycol 6000

- **Protocol**

1. To a 100ml protein solution, slowly add 150ml 50% PEG-6000 (w/v, in distilled water) while stirring gently.

2. Continue stirring for 30 - 60 minutes to allow precipitation to come to completion.

3. Centrifuge 10 minutes at 10,000 x g.

4. Decant supernatant.

5. Resuspend pellet in 1 - 2 pellet volumes of buffer.

- **Comments**

1. The solubility of a protein is reduced near its isoelectric point.

2. To remove PEG from the precipitate, an ammonium sulfate precipitation is suggested. Alternatively or in addition, ion exchange chromatography has been used (Fried and Chun, 1971). Gel filtration columns, however, have been reported to behave anomolously in the presence of elevated levels of PEG.

D. Ultrafiltration

Concentration of protein by ultrafiltration proceeds by forcing the liquid in a protein solution through a membrane which retains the protein of interest. Elevated pressure has been largely superceded by centrifugal force as the means of forcing liquid through the membrane. Thus, our description will be limited to the Amicon Centricon system as an example of ultrafiltration (Amicon Publication No. I-259C, 1986). This method is less likely to cause protein denaturation than the precipitative methods described above.

The Centricon Microconcentrator has a starting capacity of 2ml. A 2ml sample of 2mg/ml bovine serum albumin (BSA) may be concentrated to less than 50µl in a 30 minute centrifugation. By adding a new buffer to the concentrated sample and recentrifuging, buffer changes and desalting can be easily achieved (Penefsky, 1977; Christopherson, 1983). At present, two models are available, Centricon-10 and Centricon-30, with 10 kd and 30 kd average molecular weight cutoffs.

- **Equipment**

1. Centricon Microconcentrator
2. Centrifuge and rotor

- **Reagents:** None

- **Protocol**

1. Assemble the Centricon Microconcentrator and add up to 2.0ml of sample to the sample reservoir (Fig. 4.1).
2. Centrifuge at 5000 x g for 30 minutes.
3. Remove filtrate cup and apply retentate cup to the sample reservoir.
4. Invert the Centricon unit and centrifuge at 1000xg for 2 minutes.
5. The concentrated sample will be recovered in the retentate cup. A small amount of buffer may be used to wash the filter for more complete protein recovery.

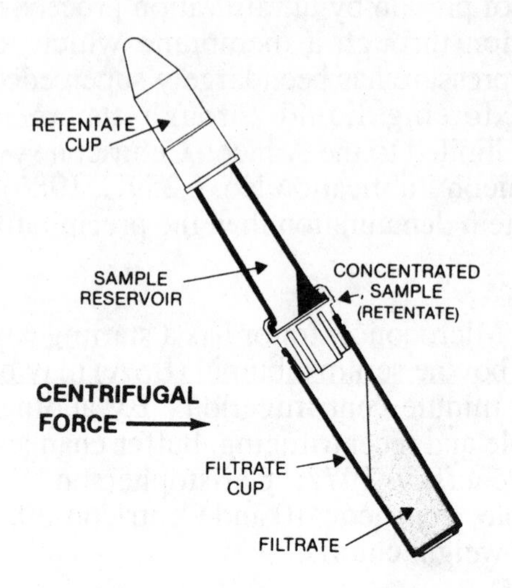

Figure 4.1. Concentration of a protein solution using the Amicon Centricon system.

• Comments

1. Centrifuge times may be decreased depending on the characteristics of the sample. Viscous solutions containing glycerol or high protein concentrations, for example, require much longer centrifugation times.

2. Recovery of proteins is typically greater than 90%.

3. The protein of interest should be 30 - 50% larger than the stated cutoff to ensure that it will be retained.

4. Different average pore sizes in different membranes may permit a slight enrichment of the protein by selective retention on the basis of protein size.

5. The Centricon Microconcentrator is designed to be used in a fixed-angle rotor. Depending on the angle of the rotor, the sample will reach a deadstop volume of 25 - 40µl beyond which no further concentration will occur. This prevents the sample from being filtered to dryness.

6. Large volumes can be accomodated in other models such as the Amicon Centriflo (7ml capacity) and the Centriprep (15ml capacity) Concentrators.

7. For certain applications, the investigator may prefer to use a pressurized and stirred cell such as the Amicon Ultrafiltration Stirred Cells, and is referred to the Amicon catalog for further details.

8. Penefsky (1977) describes a simple centrifuge column which may be used with gel filtration resin for salt removal.

E. Dialysis

Dialysis, typically used for changing the buffering solution of a protein, is also a method for concentrating protein solutions by dialysing against a hygroscopic environment (e.g. PEG, Sephadex). The protein solution is contained within a membrane which permits solute exchange with a surrounding solution and whose pore size prevents the protein from escaping.

Except for small volumes, this method based on diffusion is time-consuming, and we prefer to employ the Centricon system for preparative protein concentration and buffer exchange. However, in some cases, time is not a limiting factor and dialysis may be preferred (theory: McPhie, 1971).

- **Equipment**

 1. Beaker
 2. Magnetic stir plate and stir bar
 3. Dialysis tubing

- **Reagents**: Sephadex G-100 or G-200

- **Preparing Dialysis Tubing**

Dialysis tubing contains chemical contaminants from the manufacturing process. To remove these, it is common to boil the tubing for at least 30 minutes in 10mM sodium bicarbonate ($NaHCO_3$)/1mM EDTA. Some authors recommend more thorough treatments (see Richmond et al., 1985). Following the boiling step, the tubing should be washed extensively in distilled water and stored at 4°C in 1mM EDTA to prevent microbial contamination. Prepare new dialysis tubing every six months (Schleif and Wensink, 1981).

- **Protocol**

 1. Make two tight knots at one end of the tubing.

 2. Using a pipet or funnel (Fig. 4.2), deliver protein solution into the dialysis tubing.

 3. Tie a double knot at the other end of the tubing and place the closed dialysis bag in >10 volumes of dialysis buffer. The buffer should be stirred gently with a magnetic stir bar to improve solute exchange. Equilibrium occurs after several hours of dialysis, and the dialysis buffer may have to be changed several times until certain buffer components are sufficiently diluted.

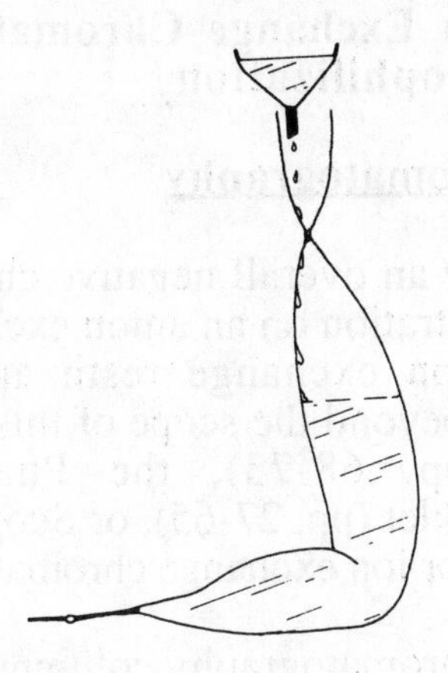

Figure 4.2. Filling dialysis tubing with a protein solution.

- **Dialysis Buffer**

For the purposes of protein concentration, the dialysis bag may be incubated in a buffer containing 20% polyethylene glycol (PEG 20,000) or 5% ethylene glycol. However, use of PEG may be accompanied by leaching of unwanted compounds into the dialysate which might require a further purification step. A more rapid method for concentrating a protein solution is to embed the dialysis sack in Sephadex G-100 or G-200 resin or Calbiochem Aquacides at 4°C. A fivefold concentration can be achieved in 3 hours by changing the resin surrounding the dialysis sack every half hour (Schleif and Wensink, 1981).

- **Comments**

1. Commercially available dialysis clips (Spectrum) are popular as an alternative to tying knots in the dialysis tubing.

2. If dialyzing against a buffer with a lower salt or organic solvent concentration than the dialysate, be aware that osmotic forces will cause water influx into the dialysis sack. Allow space for volume increases to avoid the risk of the membrane bursting.

3. Dialysis tubing generally has a 15 - 20 kd cutoff (Scopes, 1982).

4. A description of a simple microtechnique for dialyzing small volumes (10μl to 0.6ml) in an Eppendorf tube can be found in Overall (1987).

F. Briefly Noted: Ion Exchange Chromatography and Lyophilization

Ion Exchange Chromatography

Most proteins display an overall negative charge at pH 8 and thus are candidates for concentration on an anion exchange resin. The methods for preparing the ion exchange resin and carrying out column chromatography are beyond the scope of this volume. See Schleif and Wensink (1981, pp. 68-73), the Pharmacia Ion Exchange Chromatography booklet (pp. 27-65), or Scopes (1982, pp. 75-101) for detailed instructions for ion exchange chromatography.

- **Equipment** : Chromatography columns

- **Reagents**
  1. DEAE-Sephadex or DEAE-cellulose resin
  2. Tris
  3. NaCl

- **Protocol**

  1. Apply the protein solution to a DEAE column (cellulose or Sephadex) equilibrated with 20mM Tris-HCl (pH 8). 5mg of crude extract protein is the maximum that can be expected to bind per ml of anion exchange resin.
  2. Proteins bound on the column may be eluted by passing buffer of increasing ionic strength over the column. Nearly all proteins will elute with 1M NaCl.

- **Comments**

  1. A rapid method for determining a good binding pH for the protein of interest is described in the Pharmacia Ion Exchange Chromatography booklet, pp. 30-31.
  2. For an example of protein concentration using ion exchange chromatography, see Nelson (1986).

Lyophilization

Lyophilization is a commonly used method for concentration and storage of protein solutions involving sublimation of liquid from the sample in the frozen state. The reader is referred to Everse and Stolzenbach (1971) for advice and to their lyophilizer instruction manual for operating instructions.

III. Suppliers

TCA: Fluka, Merck, Serva, Sigma

Fixed *Staphylococcus aureus* cells: Bethesda Research Laboratories, Calbiochem

Protein A - Sepharose Beads: Bio-Rad, Pierce

Protein G - Sepharose Beads: Pharmacia-LKB, Pierce

Nonidet P-40: Fluka, Sigma

Microconcentrator: Amicon Centricon, Bio-Rad Unisep Cartridges

Ultrafiltration Stirred Cells: Amicon, Sartorius, Schleicher & Schuell

Dialysis Membranes: Spectrum

Sephadex G-50, G-100, or G-200 Gel Filtration Resin: Pharmacia-LKB

DEAE-Sephadex or DEAE-Cellulose Ion Exchange Resin: Pharmacia-LKB

IV. References

Amicon Publication No. I-259C. 1986. Centricon Microconcentrators for Small-Volume Concentration.

Cabib, E. and I. Polacheck. 1984. Meth. Enzymol. 104: 415-416. Protein Assay for Dilute Solutions.

Christopherson, R.I. 1983. Meth. Enzymol. 91: 278-281. Desalting Protein Solutions in a Centrifuge Column.

Everse, J. and F.E. Stolzenbach. 1971. Meth. Enzymol. 22: 33-39. Lyophilization.

Fried, M. and P.W. Chun. 1971. Meth. Enzymol. 22: 238-248. Water-Soluble Nonionic Polymers in Protein Purification.

Hames, B.D. 1981. in Gel Electrophoresis of Proteins. A Practical Approach. Hames, B.D. and D. Rickwood, eds. 290 pages. IRL Press Limited, London.

Harlow, E. and D. Lane. 1988. Antibodies: A Laboratory Manual. 726 pages. Cold Spring Harbor Laboratory, Cold Spring Harbor, New York.

Ingham, K.C. 1984. Meth. Enzymol. 104: 351-355. Protein Precipitation with Polyethylene Glycol.

Kaufman, S. 1971. Meth. Enzymol. 22: 233-238. Fractionation of Protein Mixtures with Organic Solvents.

Kessler, S.W. 1981. Meth. Enzymol. 73: 442-459. Use of Protein A-Bearing Staphylococci for the Immunoprecipitation and Isolation of Antigens from Cells.

Lerner, M.R. and J.A. Steitz. 1979. Proc. Nat. Acad. Sci. USA 76: 5495-5499. Antibodies to Small Nuclear RNAs Complexed with Proteins are Produced by Patients with Systemic Lupus Erythematosus.

McPhie, P. 1971. Meth. Enzymol. 22: 23-33. Dialysis.

Nelson, N. 1986. Meth. Enzymol. 118: 352-369. Subunit Structure and Biogenesis of ATP Synthase and Photosystem I Reaction Center.

Overall, C.M. 1987. Anal. Biochem. 165: 208-214. A Microtechnique for Dialysis of Small Volume Solutions with Quantitative Recoveries.

Penefsky, H.S. 1977. J. Biol. Chem. 252: 2891-2899. Reversible Binding of P_i by Beef Heart Mitochondrial Adenosine Triphosphatase.

Peterson, G.L. 1977. Anal. Biochem. 83: 346-356. A Simplification of the Protein Assay Method of Lowry et al. Which is More Generally Applicable.

Pharmacia Fine Chemicals. Ion Exchange Chrimatography: Principles and Methods. 71 pages.
Richmond, V.l., R. St. Denis, and E. Cohen. 1985. Anal. Biochem. 145: 343-350. Treatment of Dialysis Membranes for Simultaneous Dialysis and Concentration.
Sargent, M.G. 1987. Anal. Biochem. 163: 476-481. Fiftyfold Amplification of the Lowry Protein Assay.
Shantz, E.M. 1983. PANSORBIN *Staphylococcus aureus* Cells: Review and Bibliography of the Immunological Applications of Fixed Protein A-Bearing *Staphylococcus aureus* Cells. Calbiochem booklet. 56 pages.
Scheidtmann, K.H. 1989. pp. 109-112 in Protein Structure: A Practical Approach. T.E. Creighton, ed. 355 pages. IRL Press, Oxford.
Schleif, R.F. and P.C. Wensink. 1981. Practical Methods in Molecular Biology. Springer-Verlag, New York.
Scopes, R.K. 1982. Protein Purification. Principles and Practice. Springer-Verlag, New York.

Chapter 5

Gel Electrophoresis under Denaturing Conditions

I. SDS-Polyacrylamide Gel Electrophoresis (Linear Slab Gel)

A. Introduction

Sodium dodecyl sulfate - polyacrylamide gel electrophoresis (SDS-PAGE) is a low-cost, reproducible, and rapid method for quantifying, comparing, and characterizing proteins. This method separates proteins based primarily on their molecular weights (Laemmli, 1970). SDS binds along the polypeptide chain, and the length of the reduced SDS-protein complex is proportional to its molecular weight. The relative ease of performing SDS-PAGE along with its many applications has made it an important analytical technique in many fields of research.

Our description of SDS-PAGE will include protocols for linear slab gels, i.e., gels formed as slabs between two sheets of supporting glass. Slab gels have become more widely used than tube gels (formed with glass tube supports), since many samples can be run on the same gel, thereby providing uniformity during polymerization, staining, and destaining. For most analytical applications, the mini slab gel has gained considerable popularity due to the increased resolution and reduced amounts of time and materials required for electrophoresis. The experimental procedures and reagents in this volume have been calculated for use with a mini-gel system; however, all procedures should be readily adaptable to other systems. For certain applications, gradient gels may be useful (see Section II).

• Among the varied uses of this technique are:
1. Analysis of protein purity
2. Determination of protein molecular weight
3. Verification of protein concentration
4. Detection of proteolysis
5. Identification of immunoprecipitated proteins
6. First stage of immunoblotting (see Chapter 8)
7. Detection of protein modification
8. Separation and concentration of protein antigens for antibody production
9. Separation of radioactively labeled proteins

•Theory: Blackshear, 1984

• Sensitivity of staining:
1. Coomassie Blue: 0.1 - 1μg per band (Smith, 1984)
2. Silver Staining: 2 - 10ng per band (Giulian et al., 1983)

• Optimal Resolution Ranges (adapted from Hames, 1981)

Acrylamide Percentage	Separating Resolution
15% Gel	15 to 45 kd
12.5% Gel	15 to 60 kd
10% Gel	18 to 75 kd
7.5% Gel	30 to 120 kd
5% Gel	60 to 212 kd

• Time Required:

1. Individual Steps:

Pouring Separating Gel	60 minutes
Pouring Stacking Gel	30 minutes
Loading Samples	15 minutes
Electrophoresis	45 minutes
Staining	
Coomassie Staining	30 minutes (major bands)
Silver Staining	3 hours

2. Total Time:
   Coomassie Stained Gel: 3 hours for major bands to destain.
   Complete destaining may require 24 - 48 hours.
   Silver Stained Gel: 6 hours

B. Equipment

• Minigel apparatus
  We use the Bio-Rad Mini-Protean apparatus as a model, although all protocols should be easily adaptable to other systems. We highly recommend the minigel systems due to the savings in material and time and also because they provide high resolution protein separation.
• Power supply (capacity 200V, 500mA)
• Boiling water bath or 100°C sand bath
• Eppendorf centrifuge (optional)
• Hamilton Syringes (50µl and 100µl capacity)
• Gel dryer and high vacuum pump or water pump (optional)
• Small glass or plastic container with lid (i.e. 12 x 16 x 3cm)
• Eppendorf tubes
• Rocking or rotary shaker

C. Pouring a Gel

- **Reagents**
  1. Acrylamide, electrophoresis grade
  2. Bis-acrylamide (N, N'-methylenebisacrylamide)
  3. Tris (2-hydroxymethyl-2-methyl-1, 3-propanediol)
  4. SDS (sodium dodecyl sulfate or sodium lauryl sulfate)
  5. TEMED (N, N, N', N'-tetramethylene-ethylenediamine)
  6. Ammonium persulfate
  7. 2-mercaptoethanol
  8. Glycerol
  9. Bromophenol blue
  10. Glycine
  11. Hydrochloric acid (HCl)
  12. Dithiothreitol (DTT)

- **Stock Solutions**

  1. 2M Tris-HCl (pH 8.8), 100ml
     a. weigh out 24.2g Tris
     b. add to 50ml distilled water
     c. add concentrated HCl slowly to pH 8.8 (about 4ml) (allow solution to cool to room temperature, pH will increase)
     d. add distilled water to a total volume of 100ml

  2. 1M Tris-HCl (pH 6.8), 100ml
     a. weigh out 12.1g Tris
     b. add to 50ml distilled water
     c. add concentrated HCl slowly to pH 6.8 (about 8ml) (allow solution to cool to room temperature, pH will increase)
     d. add distilled water to a total volume of 100ml

  3. 10% (w/v) SDS, 100ml, store at room temperature
     a. weigh out 10g SDS
     b. add distilled water to a total volume of 100ml

  4. 50% (v/v) glycerol, 100ml
     a. pour 50ml 100% glycerol
     b. add 50ml distilled water

  5. 1% (w/v) bromophenol blue, 10ml
     a. weigh out 100mg bromophenol blue
     b. bring to 10ml with distilled water and stir until dissolved Filtration will remove aggregated dye.

- **Working Solutions**

1. Solution A (Acrylamide Stock Solution), 100ml
30% (w/v) acrylamide, 0.8% (w/v) bis-acrylamide

Caution: Unpolymerized acrylamide is a skin irritant and a neurotoxin. Always handle with gloves. See Safety Notes.

a. 29.2g acrylamide
b. 0.8g bis-acrylamide
Add distilled water to make 100ml and stir until completely dissolved. Work under hood and keep acrylamide solution covered with Parafilm until acrylamide powder is completely dissolved.
Can be stored for months in the refrigerator.

2. Solution B (4x Separating Gel Buffer), 100ml

a. 75ml 2M Tris-HCl (pH 8.8) ---1.5M
b. 4ml 10% SDS ---0.4%
c. 21ml H_2O
Stable for months in the refrigerator.

3. Solution C (4x Stacking Gel Buffer), 100ml

a. 50ml 1M Tris-HCl (pH 6.8) ---0.5M
b. 4ml 10% SDS ---0.4%
c. 46 ml H_2O
Stable for months in the refrigerator.

4. 10% ammonium persulfate, 5ml

a. 0.5g ammonium persulfate
b. 5ml H_2O
Stable for months in a capped tube in the refrigerator.

5. Electrophoresis Buffer, 1 liter

a. 3g Tris ---25mM
b. 14.4g glycine ---192mM
c. 1g SDS ---0.1%
d. H_2O to make 1 liter
pH should be approximately 8.3.
Can also make a 10x stock solution.
Stable indefinitely at room temperature.

6. 5x Sample Buffer, 10ml

a. 0.6ml 1M Tris-HCl (pH 6.8) ---60mM
b. 5ml 50% glycerol ---25%
c. 2ml 10% SDS ---2%
d. 0.5ml 2-mercaptoethanol ---14.4mM
e. 1ml 1% bromophenol blue ---0.1%
f. 0.9ml H_2O

Stable for weeks in the refrigerator or for months at -20°C.

• Amounts of Working Solutions to Use

1. Volumes necessary for pouring gels of different thicknesses (for two 6 x 8cm gels)

Gel Thickness	Separating	Stacking
0.5mm -	5.6ml	1.4ml
0.75mm -	8.4ml	2.1ml
1.0mm -	11.2ml	2.8ml
1.5mm -	16.8ml	4.2ml

Always prepare with a moderate excess of gel solution.

2. Calculation for X% Separating Gel

Solution A	$x/3$ ml	3.33
Solution B	2.5ml	
H_2O	$(7.5-x/3)$ml	4.17
10% Ammonium Persulfate	50µl	
TEMED	5µl (10µl if x<8%)	
--		
Total Volume	10ml	

Do not prepare until following instructions in the next section, Pouring the Separating Gel.

- **Pouring the Separating Gel**

Example of Separating Gel Preparation

Two 8% Separating Gels (6x8cmx0.75mm), 10ml
2.7ml Solution A
2.5ml Solution B
4.8ml H_2O

50µl 10% Ammonium Persulfate
5µl TEMED

Do not prepare until following numbered instructions below.

1. **Assemble gel sandwich** according to the manufacturer's instructions, in the case of commercial apparatus (e.g., Bio-Rad Mini-Gel), or according to the usage of alternative systems. For Mini-Gel, be sure that the bottom of both gel plates and spacers are perfectly flush against a flat surface before tightening clamp assembly (Fig. 5.1). A slight misalignment will result in a leak.

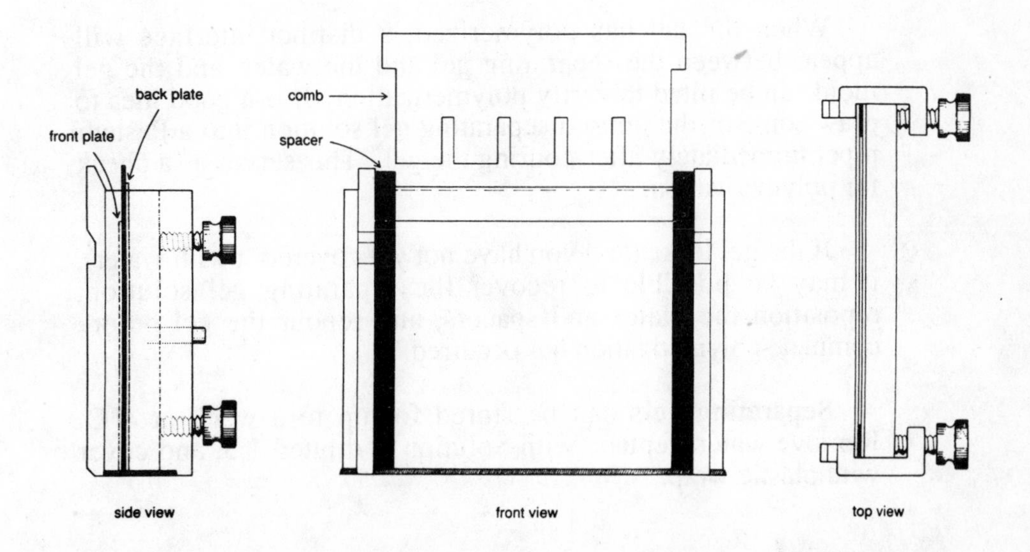

Figure 5.1. Bio-Rad Mini-Protean apparatus. Views of gel plate assembly.

2. **Combine Solutions A and B and water** in a small Erlenmeyer flask or a test tube. Acrylamide (in Solution A) is a neurotoxin, so plastic gloves should be worn at all times

3. **Add ammonium persulfate and TEMED, and mix** by swirling or inverting container gently (excessive aeration will interfere with polymerization). Work rapidly at this point because polymerization will be under way.

4. Carefully **introduce solution into gel sandwich** using a pipet. Pipet solution so that it descends along a spacer (Fig. 5.2). This minimizes the possibility of air bubbles becoming trapped within the gel.

5. When the appropriate amount of separating gel solution has been added (in the case of the Mini-Gel, about 1.5cm from top of front plate or 0.5cm below level where teeth of comb will reach, Fig. 5.3), gently **layer about 1cm of water** on top of the separating gel solution. This keeps the gel surface flat.

6. **Allow gel to polymerize** (30 - 60 minutes).

When the gel has polymerized, a distinct interface will appear between the separating gel and the water, and the gel mold can be tilted to verify polymerization. It is a good idea to draw some of the unused separating gel solution into a Pasteur pipet immediately after pouring the gel. This serves as a check for polymerization.

If the gel leaks (and you have not yet covered it with water), it may be possible to recover the separating gel solution, reposition the plates and spacers, and repour the gel before complete polymerization has occurred.

Separating gels can be stored for up to a week at 4°C. Remove water, replace with Solution B diluted 1:3, and cover with plastic wrap.

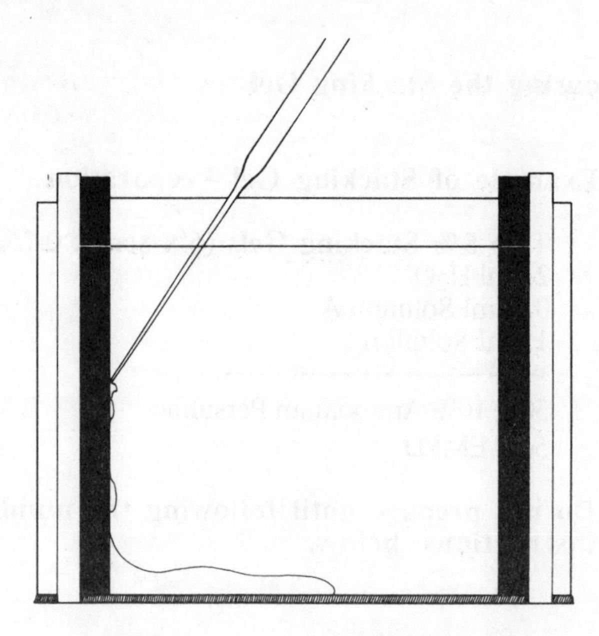

Figure 5.2. Introducing the separating gel solution into the gel sandwich.

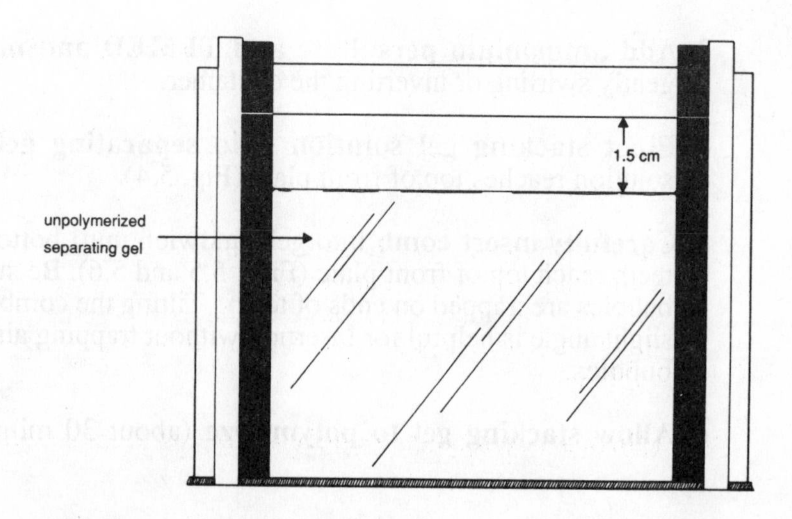

Figure 5.3. Separating gel prior to polymerization.

›uring the Stacking Gel

Example of Stacking Gel Preparation

Two 5% Stacking Gels (6 x 8cm x 0.75mm), 4ml
2.3ml H_2O
0.67ml Solution A
1.0ml Solution C

30μl 10% Ammonium Persulfate
5μl TEMED

Do not prepare until following the numbered instructions below.

1. **Pour off water** covering the separating gel. The small droplets remaining will not disturb the stacking gel.

2. **Combine Solutions A and C and water** in a small Erlenmeyer flask or a test tube.

3. **Add ammonium persulfate and TEMED and mix** by gently swirling or inverting the container.

4. **Pipet stacking gel solution onto separating gel** until solution reaches top of front plate (Fig. 5.4).

5. **Carefully insert comb** into gel sandwich until bottom of teeth reach top of front plate (Figs. 5.5 and 5.6). Be sure no bubbles are trapped on ends of teeth. Tilting the comb at a slight angle is helpful for insertion without trapping air bubbles.

6. **Allow stacking gel to polymerize** (about 30 minutes).

7. After stacking gel has polymerized, **remove comb** carefully (making sure not to tear the well ears).

8. Place gel into electrophoresis chamber. If using the Mini-Gel system, **attach both gels to electrode assembly before inserting into electrophoresis tank.**

9. **Add electrophoresis buffer to inner and outer reservoir**, making sure that both top and bottom of gel are immersed in buffer (Fig. 5.7).

Check wells for trapped air bubbles and damaged well ears. Distorted well ears can be repositioned using a Hamilton syringe.

Air bubbles clinging to bottom of gel should be removed to insure even current flow.

It is useful to rinse wells with electrophoresis buffer prior to loading in order to remove unpolymerized acrylamide and any contaminants.

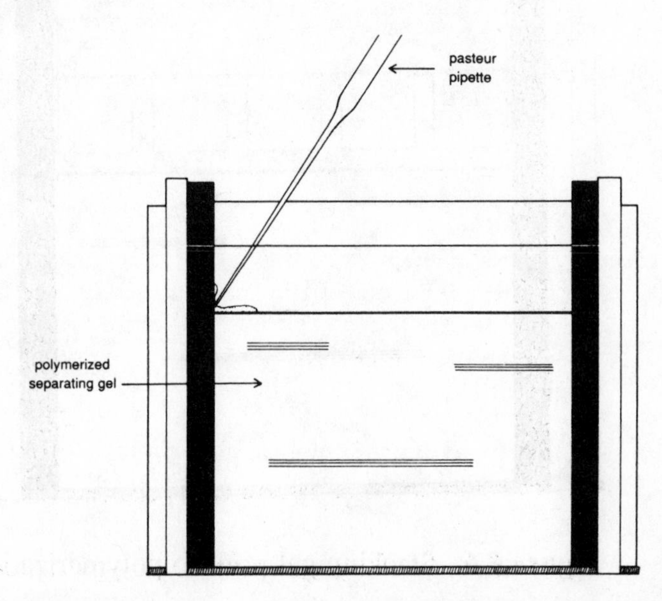

Figure 5.4. Introducing the stacking gel solution into the gel sandwich.

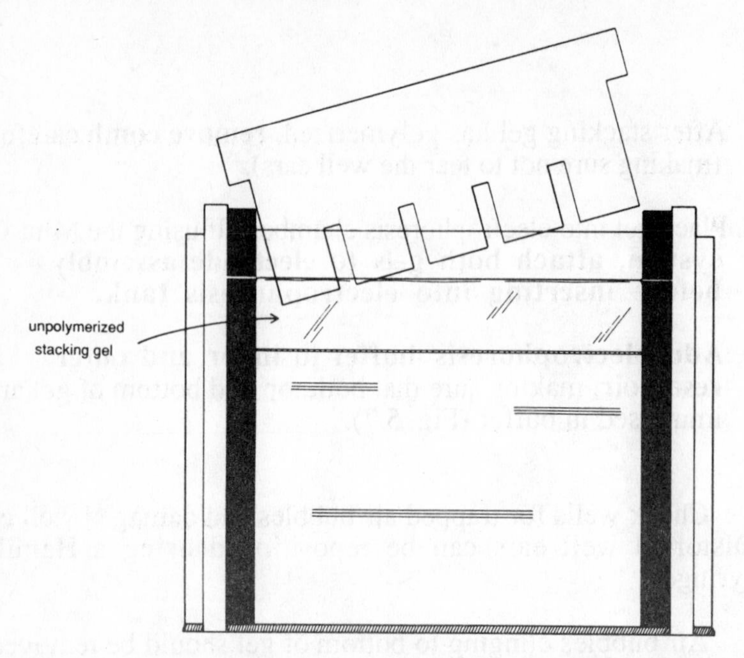

Figure 5.5. Inserting the sample-well comb into the stacking gel.

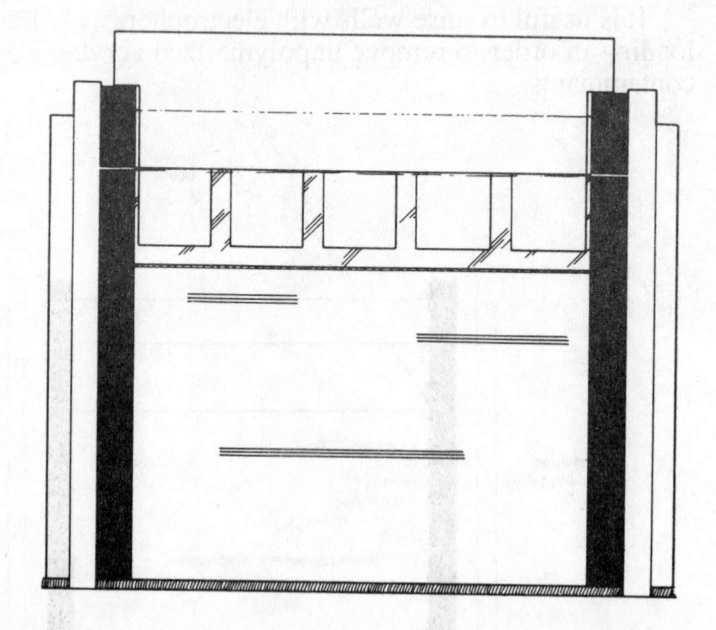

Figure 5.6. Stacking gel prior to polymerization.

Figure 5.7. Mini-Protean electrophoresis chamber.

• Comments

1. Reagent solutions are stored at 4°C and generally do not need to be warmed to room temperature before mixing and pouring gels.

2. Possible reasons that a gel does not polymerize:

   a. Not enough ammonium persulfate or TEMED. Increase volumes of catalysts.
   b. Poor quality reagents. Use electrophoresis grade reagents.
   c. Ammonium persulfate or TEMED are inactive. Prepare or purchase fresh stock.
   d. Temperature is too low. Cast gel at room temperature.

   For a more detailed troubleshooting guide, consult BioRad Bulletin 1156.

3. Polymerization rate can be most easily adjusted by altering the amounts of polymerization catalysts used.

4. Degassing the acrylamide solution will lead to more rapid polymerization. We have found this step to be cumbersome and unnecessary in most applications.

5. Gel cracking during polymerization (especially high percentage gels) is likely due to excess heat generation. Use cooled reagents (Hames, 1981).

6. If electrophoresis is to be carried out at 4°C, lithium dodecyl sulfate (LiDS) should be substituted for SDS. LiDS does not precipitate at low temperatures.

7. Sharpness of protein bands may vary according to the brand and grade of SDS used. See Rothe and Maurer (1986, pp. 106-108) for an analysis and discussion of SDS purity from different commercial sources.

8. For a polyacrylamide gel system which separates proteins from 50 kd to 500 kd, see Perret et al. (1983).

D. Preparing and Loading Samples

• **Capacity per Well** (Mini-Gel System)

Gel Thickness	1 Well	5 Wells	10 Wells	15 Wells
0.5mm	0.7ml	45μl	16μl	9μl
0.75mm	1.0ml	68μl	24μl	14μl
1.0mm	1.4ml	90μl	32μl	18μl
1.5mm	2.1ml	135μl	48μl	27μl

• **Steps**

1. **Combine protein sample and 5x Sample Buffer** (i.e. 20μl + 5μl) in an Eppendorf tube.
2. **Heat at 100°C for 2 - 10 minutes.**
3. **Spin down protein solution for 1 second** in microfuge - longer if large quantities of debris are present.
4. **Introduce sample solution into well** using a Hamilton syringe (Fig. 5.8). Layer protein solution on bottom of well and raise syringe tip as dye level rises. Be careful to avoid introducing air bubbles as this may allow some of sample to be carried to adjacent well.

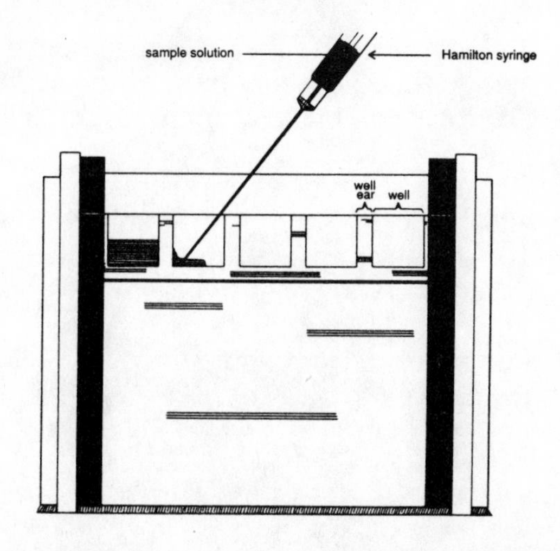

Figure 5.8. Introducing protein solution into sample well.

Rinse syringe thoroughly with electrode buffer or water before loading different samples.

Include molecular weight standards in one or both outside wells (see Section I. I. of this chapter).

Gels can also be loaded before inserting into the electrophoresis tank, but subsequent filling of the tank with electrophoresis buffer may cause some convection currents, resulting in cross-contamination of wells.

Although the stacking effect should reduce variations due to differences in loading volumes, it is recommended that equal sample loading volumes be used for analytical work.

To facilitate visualizing the gel wells, add bromophenol blue to a final concentration of 1mg/ml in the stacking gel (Smith et al., 1988).

• Comments

1. Minimum protein loading per well (single protein band): 0.1μg (Coomassie staining) to 2ng (silver staining, Giulian et al., 1983).

2. Maximum protein loading per well (mixture of proteins): 20 - 40μg.

3. Precipitation of protein in sample buffer may be due to denatured protein, too litle SDS, too little reducing agent, overly acidic conditions, or presence of potassium which precipitates SDS (Hames, 1981).

4. During sample preparation, if the loading mix turns yellow, the solution has become too acidic: add NaOH until the solution turns blue; otherwise, the protein sample may migrate anomalously.

5. To avoid edge effects, add 1x sample buffer to unused wells.

6. If the sample does not sink to the bottom of the well, either there is insufficient glycerol in the sample buffer or the comb did not fit snugly, leaving polymerized acrylamide deposits in the well (Hames, 1981).

7. If boiled samples are not centrifuged prior to loading, streaking of protein bands may result. Streaking may arise from precipitation due to accumulation of very high protein concentrations in the stacking gel. The aggregated proteins are thought to dissolve slowly during the course of electrophoresis. Diluting the protein sample may help (Hames, 1981).

8. Boiled samples in sample buffer are usually stable for weeks if stored frozen (-20ºC), although repeated freeze-thawing may lead to protein degradation.

9. If boiled samples are stored in a freezer, the samples should be warmed prior to loading because SDS precipitates out of solution (Hames, 1981).

10. When loading wells, take care not to overfill them. Contaminating the adjacent well can create troublesome artifacts.

E. Running a Gel

- **Steps**

1. **Attach electrode plugs** to proper electrodes. Current should flow towards the anode.

2. **Turn on power supply** to 200V (constant voltage; amperage will be about 100mA at start, 60mA at end of electrophoresis for two 0.75mm gels; 110mA at start, 80mA at end for two 1.5mm gels).

3. The dye front should migrate to 1cm from the bottom of the gel in 30 - 40 min for two 0.75mm gels (40 - 50 min for 1.5mm gels).

Gels will become quite warm during electrophoresis, although this does not affect separation. Lower voltage differences result in less heat generation, slower migration times, and possibly slightly decreased resolution.

After electrophoresis, gels may stand for a few hours before staining without harm except for gels with low percentage acrylamide in which protein will start to diffuse.

The high electrical current used in gel electrophoresis is very dangerous. **Never disconnect electrodes before first turning off the power source.** If using an electrophoresis apparatus which is not completely shielded from the environment, always leave a clearly visible sign warning that electrophoresis is in progress.

4. **Turn off power supply.**

5. **Remove electrode plugs** from electrodes.

6. **Remove gel plates** from electrode assembly.

7. Carefully remove a spacer, and, inserting the spacer in one corner between the plates, gently **pry apart the gel plates.** The gel will stick to one of the plates.

- **Comments**

1. Only constant voltage conditions give constant protein mobility during electrophoresis (Hames, 1981).

2. Uneven band migration is often due to uneven electrical current flow. Make sure that electrophoresis buffer in both upper and lower chambers is making good contact with the gel. Other possible reasons include overloading of protein in wells, high salt content of the sample or effects at wells on the end of the gel. "Smiling," or reduced mobility of samples at the edge of the gel, may be due to the center of the gel being hotter than the sides; decreasing the power setting may help.

F. Staining a Gel with Coomassie Blue

This method of staining can detect as little as 0.1μg of protein in a single band (Fig. 5.9). Generally a choice is made between using Coomassie Blue or Silver Stain (see Section G) depending on sensitivity desired.

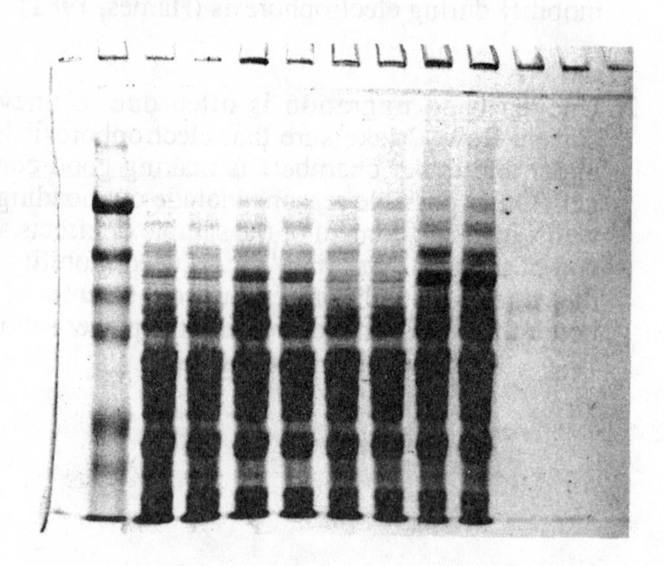

Figure 5.9. Gel stained with Coomassie blue. The example presented represents a control experiment to verify that similar concentrations of protein were prepared in extracts of yeast strains grown under different conditions. The experiment was performed with a 1.5mm gel in 8% acrylamide subjected to 200V for 40 min. Protein concentrations were chosen to give 10μg per well.

- **Reagents**
  1. Coomassie blue R-250
  2. Methanol - CH_3OH
  3. Glacial acetic acid - CH_3COOH

- **Stock Solutions**

1. Coomassie Gel Stain, 1 liter
   1.0g Coomassie Blue R-250
   450ml methanol
   450ml H_2O
   100ml glacial acetic acid
2. Coomassie Gel Destain, 1 liter
   100ml methanol
   100ml glacial acetic acid
   800ml H_2O

- **Staining Procedure**

1. Wearing gloves to prevent transfer of fingerprints to the gel, **pick up the gel and transfer it to a small container** (taking care not to tear the gel) containing a small amount of Coomassie Stain (20ml is sufficient), or gently agitate the glass plate in stain solution until gel separates from plate.
2. **Agitate for 5 - 10 minutes** for 0.75mm, 10 - 20min for 1.5mm gel on slow rotary or rocking shaker. Cover container with lid or plastic wrap during staining and destaining.
3. **Pour out stain** (can be reused several times, but it is fairly inexpensive so we generally discard it) and rinse the gel with a few changes of water. Use gloves to avoid staining hands.
4. **Add Coomassie Destain** (about 50ml). Strong bands are visible immediately on a light box, and the gel is largely destained within an hour. Used destain can be washed down the sink with ample water.
5. To destain completely, change destain solution and agitate overnight. 1 - 2cm of yarn or a piece of styrofoam can be added to absorb Coomassie stain which diffuses from the gel.

• Comments

1. The staining and destaining times given are minimum incubation times, but staining overnight will only require longer destaining. If staining appears to be incomplete after destaining, gel can be restained.

2. For more fragile gels (low percentage polyacrylamide), transfer gel to staining container by placing a sheet of Whatman paper over the gel while still on the glass plate, then lift paper off with gel clinging to it and wash gel from paper in the staining container.

3. High concentrations of SDS may interfere with Coomassie staining (Hames, 1981).

4. Uneven staining is most likely due to incomplete penetration of the dye either because not enough dye was added or because agitation was insufficient.

5. Nonspecific Coomassie staining may be due to deposition of undissolved dye. Filter the dye solution.

6. Dyes tend to be attracted by positively charged groups (Lys, Arg), thus basic proteins tend to stain more strongly and some acid proteins have escaped detection (Scopes, 1982).

7. If Coomassie staining is not sensitive enough, the gel can be rinsed and subsequently silver stained. See silver staining comments below.

8. A single container may be used for staining and destaining. Reagent grade methanol works well for removing residual Coomassie stain from the container.

G. Silver Staining a Gel

This method of staining can detect as little as 2ng of protein in a single band (Fig. 5.10).

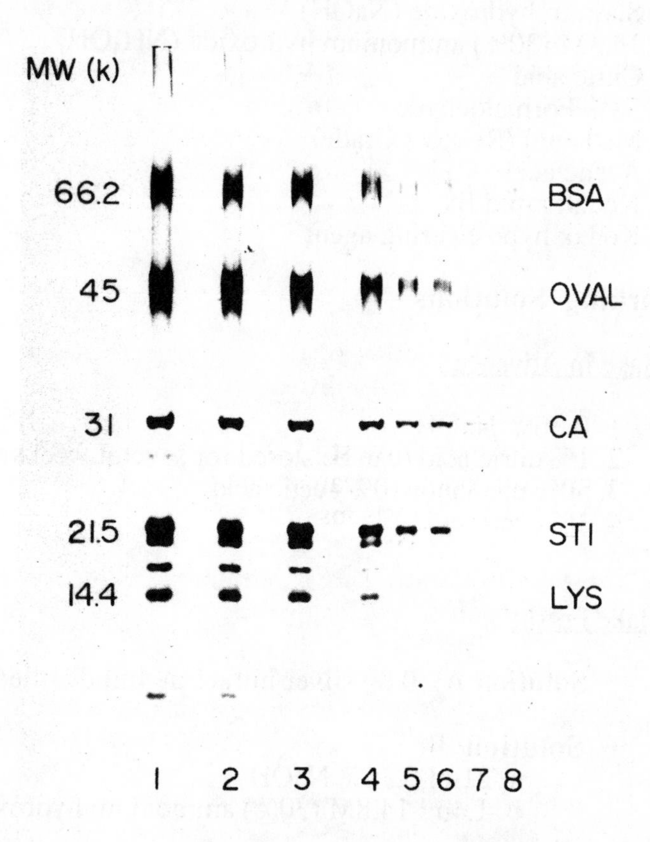

Figure 5.10. Gel stained with silver. Protein samples from the Bio-Rad low molecular weight kit were examined at dilutions of a stock solution with the proteins at the following concentrations (μg/ml): bovine serum albumin (1.2), ovalbumin (1.7), carbonic anhydrase (1.9), soybean trypsin inhibitor (1.8), and lysozyme (1.4). Lanes 1-8 were loaded with 2.0μl of the stock solution at the following dilutions: 1 (1:40), 2 (1:53), 3 (1:80), 4 (1:160), 5 (1:320), 6 (1:400), 7 (1:1600), and 8 (1:4000), resulting in samples in the range of 0.6 to 96 ng. Although the sensitivity varies somewhat among different proteins, generally 2 ng could be readily detected. Results from Giulian et al. (1983) kindly provided by G.G. Giulian.

- **Reagents**

1. Silver nitrate ($AgNO_3$) (Note: Different silver nitrate batches may have different sensitivities.)
2. Sodium hydroxide (NaOH)
3. 14.8M (30%) ammonium hydroxide (NH_4OH)
4. Citric acid
5. 38% Formaldehyde
6. Methanol (Reagent Grade)
7. Acetic acid
8. Kodak rapid fix
9. Kodak hypo clearing agent

- **Working Solutions**

Make In Advance:

1. 0.36% NaOH
2. 1% citric acid (can be stored for several weeks)
3. 50% methanol/10% acetic acid
4. 1% acetic acid

Make Fresh:

5. **Solution A**: 0.8g silver nitrate in 4ml distilled H_2O

6. **Solution B**:
   a. 21ml 0.36% NaOH
   b. 1.4ml 14.8M (30%) ammonium hydroxide

7. **Solution C**:
   Add Solution A to Solution B dropwise with constant vigorous stirring, allowing brown precipitate to clear. Add water to 100ml. Use within 15 minutes.

8. **Solution D**:
   Mix 0.5ml 1% citric acid with 50µl 38% formaldehyde, add water to 100ml. Solution must be fresh.

• **Staining**

1. Wearing gloves, pick up the gel and transfer it to a small container. **Soak gel in 50% methanol/10% acetic acid for at least 1 hour** with 2 - 3 changes of methanol/acetic acid.

2. **Rinse 30 minutes with water,** with at least 3 changes.

3. **Prepare Solutions A, B, then C.**

4. **Remove gel to a clean container and stain in Solution C for 15 minutes** with gentle, constant agitation.

5. **Rinse gel twice in deionized water, then soak 2 minutes** with gentle agitation.

6. **Prepare Solution D.**

7. **Remove gel to a clean container and develop by washing gel in Solution D.** Bands should appear in less than 10 minutes or else change Solution D. If a pale yellow background appears, reaction should be stopped.

8. **Stop development by rinsing in 1% acetic acid.**

9. **Wash gel in water for at least 1 hour** with at least three changes of water.

10. If protein deposits are too dark, destain gel with Kodak Rapid Fix or Kodak Unifix. Stop destain with Kodak hypo clearing agent such as Orbit. Then wash in 50% methanol/10% acetic acid.

11. Store gel in water or dry gel (see Section H, Drying a Gel).

• Notes on Staining Steps

Step 1:
- High quality methanol leads to poor staining; therefore, use reagent grade methanol.
- Initial methanol soak can be done for weeks.
- A more rapid method involves soaking the gel for 30 minutes in 70% PEG 2000 and proceeding directly to step 3 (Ohsawa and Ebata, 1983).
- Agarose-containing gels are more efficiently fixed in 20% TCA.

Step 4: **Solution C becomes highly explosive when dried.** Collect in a bottle and add an equal volume of 1M HCl to precipitate $AgCl_2$ (Gooderham, 1984). Silver chloride can be washed down drain with ample cold water.

Step 7: 5 - 10% methanol can be used to slow down development.

Step 10: Destaining in Rapid Fix can also be slowed with 10% methanol.

• Comments

1. If a gel is understained with Coomassie blue, simply rinse very well with methanol and continue with step 2 of the silver stain protocol. Acetic acid interferes with silver staining, so be sure that it is completely washed out of the gel. Conversely, if silver staining overstains the gel, it can be destained with Rapid Fix and restained with Coomassie blue.

2. Different proteins have a nonlinear response to silver staining and basic proteins stain especially poorly. Thus, **estimating stoichiometries should be avoided with silver stained gels**.

3. Slow agitation (40 - 60 rpm) is important during incubations.

4. Surface artifacts can be caused by pressure, fingerprints, and surface drying. Gloves should be worn at all times when handling the gel. Handling of the gel may be limited by use of a water pump to remove solutions after incubations.

5. A uniform dark background may be due to impurities in the water (Bio-Rad Bulletin 1089). Deionized water with conductivity <1μmho is required in all solutions.

6. If a gray or brown precipitate appears as dust, smudges, or swirling on the gel surface, this may be due to insufficient washing during steps 2, 5, or 9, or to low temperatures (Bio-Rad Bulletin 1089).

7. High silver staining background may be due to impurities in the acrylamide (Bio-Rad Bulletin 1156).

8. Interfering substances such as glycerol, urea, glycine, Triton X-100, and ampholytes should be removed in the initial methanol wash (Wray et al., 1981).

9. Some variability in silver staining may be due to temperature fluctuations if the protocol is carried out at room temperature. Using a constant temperature water bath may resolve such a problem (Allen et al., 1984).

10. Silver staining may not work when the protein sample contains nucleic acids or metals. Changes in fixing and washing the gel prior to staining should help (Wedrychowski et al., 1986).

11. SDS gels in which 2-mercaptoethanol is used may develop 2 horizontal lines at 60 kd and 67 kd. These can be eliminated by using less 2-mercaptoethanol (Marshall and Williams, 1984).

12. Glutaraldehyde pretreatment has been shown to enhance staining for various proteins by up to 40-fold (Dion and Pomenti, 1983). To cross-link proteins with glutaraldehyde, wash for 30 minutes with 10% glutaraldehyde in the hood after step 1, followed by a 2 hour continuous water wash (Giulian et al., 1983). Gels can be stored for several weeks in distilled water following this step. 25% glutaraldehyde stock should be stored in the refrigerator.

13. An alternative silver staining method has been described by Merril et al. (1984).

14. Glassware must be extremely clean. The following treatments of glassware have been used successfully: acid washing, NEN Count-Off, and Malinckrodt Chem-Solv.

15. Photographs should be taken as soon as possible because of color changes and increased background with time (Gooderham, 1984).

16. ^{125}I, ^{32}P, ^{35}S, ^{14}C autoradiography after silver staining is fine, but not ^{3}H fluorography because the stain will absorb most of the emissions from the isotope (Gooderham, 1984). Destaining the gel prior to fluorography will eliminate quenching.

H. Drying a Gel

• **Materials**

- Whatman 3MM paper
- Acetate sheet (the kind used for transparencies) or standard plastic kitchen wrap

• **Protocol**

1. **Place gel upside down** on a clean surface (glass plate or laboratory bench).
2. Using water as a lubricant, adjust well ears to be upright.
3. **Cover gel with a 10 x 12cm piece of Whatman 3MM paper** and lift the 3MM paper with the gel off of the glass plate or benchtop. The gel should stick to the paper.
4. **Cover the front of the gel with an acetate sheet** or plastic wrap, taking care not to trap air bubbles, which can lead to gel cracking. It is useful to roll out air bubbles, using a test tube as a rolling pin.
5. **Place Whatman paper on gel dryer, turn on heat** (Smith, 1984, recommends 60°C) **and suction, and cover with sealing gasket.**

• **Comments**

1. We find that a 0.75mm gel is dry in about 1 hour and a 1.5mm gel dries in about 2 hours, but this depends on the pump or aspirator you are using.
2. Use paper >1mm thick to prevent curling (Smith, 1984).
3. Gels can also be dried at room temperature and atmospheric pressure behind sheets of cellophane (Giulian et al., 1983 and Smith, 1984). Gels dried behind transparent sheets are useful for densitometry.
4. Cracking of a gel during drying, especially common for high percentage acrylamide gels, may be due to release of the vacuum before the gel is completely dry (Hames, 1981). Soaking the gel overnight in 5% glycerol is often recommended to decrease the risk of cracking during drying. For other strategies to avoid gel cracking, see Smith (1984).

I. General Discussion

- Examples of abnormal migration (Weber et al., 1971):

  1. Incomplete SDS binding leads to reduced electrophoretic mobilities and occurs in the following cases:
     a. proteins which are not completely reduced;
     b. chemically cross-linked proteins;
     c. glycoproteins, and succinylated, maleylated, or acidic proteins.

  2. Proteins whose intrinsic net charge makes a significant contribution to the total net charge after SDS binding, such as histones, will have deviant relative mobilities.

  3. Proteins of molecular weight below 15 kd begin to behave unusually in SDS-PAGE, due to a charge to mass ratio that is different from larger proteins and also due to the changed properties of small particles migrating in a gel. SDS-urea gels may partially overcome these problems (see section III. below).

- If doublets are observed where a single species is expected, it is possible that a portion of the protein sample may not be completely reduced. Increase the 2-mercaptoethanol concentration in the sample buffer. It is also reasonable to suspect proteolysis.

- Very hydrophobic proteins (for example *E. coli* lactose permease) may bind excess SDS and display high electrophoretic mobility. Higher acrylamide concentrations provide a more accurate molecular weight (See and Jackowski, 1989). Detergent-extracted proteins may require the presence of the detergent to remain soluble or active. Electrophoresis in the presence of the detergent may provide the most satisfactory solution, although detergents with a large micelle size such as Triton X-100 (see Chapter 1, Table 2) may penetrate poorly into the polyacrylamide gel (Rothe and Maurer, 1986, pp. 112-117).

- High concentrations of cations in the sample may cause precipitation of SDS and should be removed prior to addition of sample buffer.

- Some proteins need to be rapidly fixed or else they diffuse out of the gel (Wilson, 1983). Also, when using ultrathin gels (<0.5mm), fixing prior to staining is necessary. 20% TCA will also act as a fixative (Allen et al., 1984).

- Ammonium persulfate may interfere with enzyme activity. Pre-electrophoresis or substituting riboflavin may help. See Blackshear (1984).

- SDS often inhibits enzyme activity, but various strategies exist for recovering active enzymes from SDS-polyacrylamide gels. See Section IV. D. For an enzyme assay following SDS-PAGE, see Spanos and Huebscher (1983).

- For cleaning electrophoresis apparatus and plates, mild dishwashing detergent is generally adequate. Chromic acid washing followed by water and ethanol rinses is recommended by various investigators, but we obtain satisfactory results using only mild detergents.

- Sealing gel plates: The Bio-Rad Mini-Protean apparatus requires no sealing of the gel plates. If using a gel plate system which requires sealing, we recommend coating the mounted gel plates with a heated 1.5% agarose solution.

- Low percentage acrylamide gels (<3%) may be reinforced with 0.5% agarose (Hames, 1981).

- For gels with acrylamide concentrations above 12%, a higher acrylamide:bis ratio is recommended. See Blackshear (1984).

- %T and %C are terms which are often used to describe the acrylamide composition of the gel. %T refers to the total acrylamide content (w/v) while %C is the ratio of crosslinking reagent (i.e. bis-acrylamide) to acrylamide monomer (w/w). The following formulas apply:

$$\%T = \frac{\text{Acrylamide (g)} + \text{Bis (g)}}{\text{Volume (ml)}} \times 100\%$$

$$\%C = \frac{\text{Bis (g)}}{\text{Acrylamide (g)} + \text{Bis (g)}} \times 100\%$$

(from Pharmacia Publication)

- Minislab gels have become the system of choice for most electrophoretical applications due to their increased resolution and shorter running and staining times. Larger gels may, however, be useful for certain purposes such as providing greater physical separation between protein bands for autoradiography (Schleif and Wensink, 1981). The Pharmacia Phastsystem represents a milestone in reduction of size and time needed for gel electrophoresis.

- Protein stoichiometries can be determined from Coomassie stained gels by scanning densitometry and integration (see section VI. below). Various caveats apply; see Weber et al. (1971).

- A simple test for enzymatic proteolysis is to add the sample buffer to two identical protein solutions, and boil one but not the other protein sample (Weber et al., 1971).

- For removal of nucleic acids, see Sinclair and Rickwood (1981) or Schleif and Wensink (1981).

- Preparation of size standards: 2μg of a purified protein standard per lane is adequate. If preparing a mixture of standard proteins, scale up accordingly (i.e., for a mixture of 5 standard proteins, 10μg of the size standard solution should be loaded per well). Making aliquots for each gel will help in preventing protein degradation due to repeated freeze-thawing. Prestained molecular weight markers are extremely useful, but be aware that dye binding increases the molecular weight of the markers.

J. Safety Notes

- **Acrylamide:** Extremely toxic, causing central nervous system paralysis. Can be absorbed through unbroken skin. If skin comes in contact with acrylamide powder or solution, wash with soap and much water. Unpolymerized acrylamide should be polymerized with excess catalyst and disposed of with solid waste (Merck Index and BioRad Bulletin 1156).

- **Ammonium Persulfate:** Dispose of by diluting with water (BioRad Bulletin 1156).

- **TEMED:** Store refrigerated in a dark bottle (Hames, 1981). Significant loss of activity after 10 - 12 months (BioRad Bulletin 1156).

- **Silver Nitrate:** Poisonous and a skin irritant (Merck Index).

- **Formaldehyde:** Vapors are very irritating. Store well-closed in a moderately warm environment (Merck Index).

II. Gradient Gels

A. Introduction

The use of a gradient of acrylamide for gel electrophoresis has two advantages over the use of a linear gel. An increased protein sieving effect at higher acrylamide concentrations leads to the formation of sharper bands. The gradient also permits the separation of a larger range of protein molecular weights in a single gel (from 15 kd to 200 kd on a 5 - 20% gel or 13 kd to 950 kd on a 3 - 30% gel). The percentages of acrylamide used in the gradient may be adjusted according to the need (Hames, 1981; Walker, 1981).

B. Equipment

- Gradient Maker (Hoefer, Pharmacia-LKB)
- Peristaltic Pump (Baxter, Fisher, Pharmacia-LKB)
- Tubing (Baxter, Fisher)

C. Preparing Gels

- **Working Solutions**

1. Solution A - 30% acrylamide, 0.8% bis-acrylamide (Section I.C.)
2. Solution B - 1.5M Tris-HCl (pH 8.8), 0.4% SDS (Section I.C.)
3. Solution C - 0.5M Tris-HCl (pH 6.8), 0.4% SDS (Section I.C.)

- **Amounts of Solutions for a 5 - 20% Separating Gel**

5%	20%	
1.67ml	6.67ml	Solution A
2.5ml	2.5ml	Solution B
5.8ml	-	H_2O
-	1.5g	sucrose (adds 0.8ml to volume)
50μl	50μl	10% Ammonium Persulfate
5μl	5μl	TEMED

• Instructions for Forming Separating Gel

The only significant difference in pouring a gradient acrylamide gel as opposed to a linear gel is the use of a gradient maker. A gradient maker (Fig. 5.11) mixes the high and low concentration acrylamide solutions just prior to pouring the separating gel mixture into the gel sandwich. In the absence of a peristaltic pump, the gel solution may be introduced into the gel sandwich by gravity. Sucrose is also added to the high concentration acrylamide solution to stabilize gradient formation.

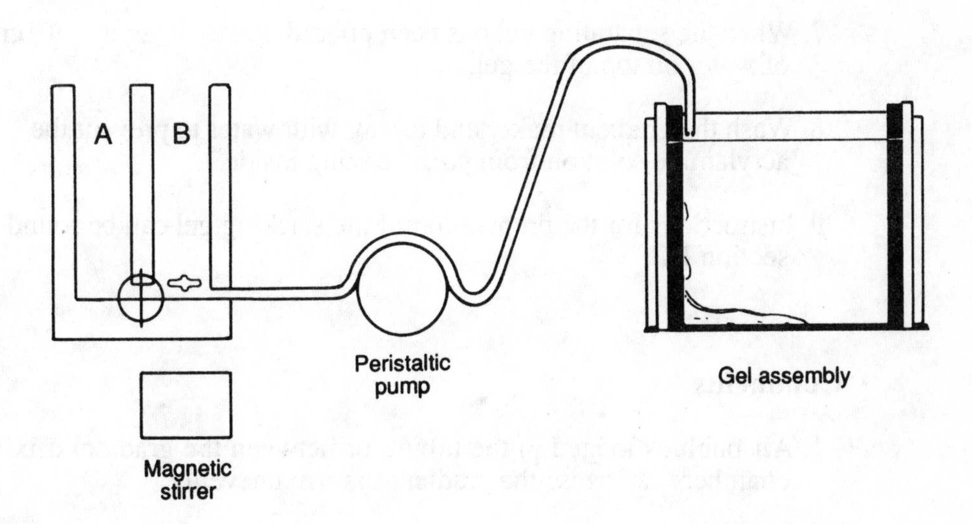

Figure 5.11. Apparatus for pouring gradient gels.

1. Prepare the gel sandwich and set up the gradient maker as shown in Fig. 5.11. The magnetic stirrer should be isolated from the gradient maker so as not to heat the mixing chambers (i.e., with a sheet of styrofoam). Calibrate the peristaltic pump to a flow rate of approximately 3 ml/min.

2. Prepare the separating gel solutions without adding TEMED.

3. Add TEMED to the separating gel solutions and mix gently. Work rapidly at this point because polymerization will be under way.

4. Transfer the appropriate volume of each solution into the mixing chambers (i.e., for a 10ml gel, add 5ml into each mixing chamber). The high concentration (20%) acrylamide solution should be added to chamber B.

5. Turn on the magnetic stirrer and open the connection between the two chambers.

6. Turn on the peristaltic pump and allow gradient gel to form.

7. When the separating gel has been poured, gently layer about 1cm of water on top of the gel.

8. Wash the gradient maker and tubing with water to prevent the acrylamide solution from polymerizing inside.

9. Instructions for the preparation of the stacking gel can be found in section I.C.

- **Comments**

1. Air bubbles lodged in the tubing or between the gradient mixing chambers can cause the gradient to form unevenly.

2. Gel solutions may be chilled to allow more time for pouring the gradient before polymerization occurs.

3. Riboflavin may be substituted for ammonium persulfate to allow more time for pouring the gradient before polymerization occurs. A final riboflavin concentration of 0.0005% (from a 0.004% stock solution) may be used, and polymerization is initiated by exposure to daylight or a white or blue fluorescent lamp.

Consult: Hames (1981) or Walker (1984).

III. SDS-Urea Gels

A. Introduction

The mobilities of small proteins in SDS-PAGE may no longer be proportional to their molecular weight when the protein charge properties become significant relative to the mass. SDS-urea PAGE is often used in these cases (Swank and Munkres, 1971; Schleif and Wensink, 1981). In addition, SDS-urea PAGE may be useful with proteins such as immunoprecipitates and membrane proteins, which are not soluble at low ionic strengths.

Aside from the gel composition described below, all procedures are as described for SDS-PAGE (Section I.).

B. Preparing Gels

- **Working Solutions**
  1. Solution A - 30% acrylamide, 0.8% bis-acrylamide (Section I.C.)
  2. Solution B - 1.5M Tris-HCl (pH 8.8), 0.4% SDS (Section I.C.)
  3. Solution C - 0.5M Tris-HCl (pH 6.8), 0.4% SDS (Section I.C.)

- **Amounts of Solutions**

For X% Separating Gel (10ml)		**X% Stacking Gel** (4ml)	
Solution A	$^{x}/_{3}$ml		$^{x}/_{7.5}$ml
Solution B	2.5ml	Solution C	1ml
urea	4.8g (=3.6ml)		1.9g (=1.4ml)
H_2O	3.9 - $^{x}/_{3}$ml		1.6 - $^{x}/_{7.5}$ml
10% Ammonium Persulfate	50μl		30μl
TEMED	5μl		5μl

Note: These formulas for separating and stacking gels are valid only up to an acrylamide concentration of 12%. For higher acrylamide concentrations, urea should be included in Solutions A and B. See Schleif and Wensink (1981).

• **Example**

8% Separating Gel (10ml)		**5% Stacking Gel** (4ml)	
Solution A	2.7ml		0.67ml
Solution B	2.5ml	Solution C	1.0ml
urea	4.8g		1.9g
H_2O	1.2ml		0.93ml
10% Am. Persulfate	50µl		30µl
TEMED	5µl		5µl

Electrophoresis conditions and buffer solutions as well as staining and destaining solutions are the same as described for SDS-PAGE (Section I.). 5x Sample buffer should be made to contain 8M urea.

IV. Other Methods

A. Detection of Radiolabeled Samples

The use of radioactively labeled proteins for gel electrophoresis has extended the limits of protein detection. Radioactivity can be measured by autoradiography (exposure of the gel to X-ray film) or by cutting gel slices and counting them in a scintillation counter.

Gel slicing is a more quantitative assay, although it suffers from poor resolution. Gels may be sliced with a razor blade, and commercial gel slicers are also available (Bio-Rad, Hoefer). For good liquid scintillation counting efficiency, proteins should first be eluted from the gel slice. A simple procedure for dissolving gel slices with H_2O_2 is described by Hames (1981, pp. 56-58).

Autoradiography may involve direct exposure of the X-ray film to the gel (sealed in plastic wrap) or may be enhanced more than tenfold by the use of an intensifying screen behind the film (called indirect autoradiography). The sensitivity of detection for ^{14}C, ^{32}P, and ^{35}S is enhanced approximately 15-fold by the use of fluorographic methods. Among the more popular of these is gel impregnation with a fluor such as EN^{3}HANCE (New England Nuclear). Fluorography is necessary for detection of ^3H. Methods for autoradiography and fluorography are described in the references listed below.

Consult: Bonner (1984), Hames (1981), Harlow and Lane (1988), Roberts (1985), or Waterborg and Matthews (1984).

B. Molecular Weight Determination

SDS-PAGE is frequently used to determine the molecular weight of a protein since protein migration is generally proportional to the mass of the protein. A standard curve is generated with proteins of known molecular weight, and the molecular weight of the protein of interest can be extrapolated from this curve.

• Protocol

1. Following gel electrophoresis and staining, measure the distance of migration of the proteins as well as that of the tracking dye (bromophenol blue). Distance of migration is measured from the beginning of the separating gel to the leading edge of a protein band.

2. Calculate R_f values

$$R_f = \frac{\text{distance of protein migration}}{\text{distance of tracking dye migration}}$$

3. Plot the $\log_{10}$ of the known protein molecular weights as a function of their R_f. The area in the middle of the gel should yield a straight line (Fig. 5.12).

4. Read molecular weight of the unknown protein from the graph based on its R_f value.

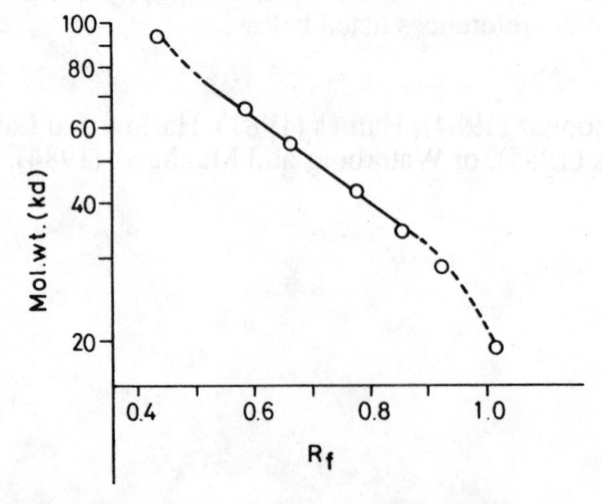

Figure 5.12. Semilogarithmic graph of molecular weight versus relative mobility. Modified from See and Jackowski (1989).

• Comments

1. Calculations of R_f values should be made from proteins separated on the same slab gel. This eliminates variability due to acrylamide concentration and electrophoretic conditions.

2. Molecular weight determinations based solely on SDS-PAGE may be misleading, since some proteins migrate anomolously.

3. For a careful molecular weight determination, it is advisable to run gels of at least two different acrylamide concentrations.

4. A new standard curve must be generated for each gel.

5. Protein size standards are available commercially. A table of selected protein molecular weights and isoelectric points is provided in Appendix 2.

6. Be careful to make appropriate molecular weight corrections for prestained molecular weight markers. Dye binding to standard proteins increases their molecular weight. Consult the specification sheet accompanying the prestained markers.

7. Gels may be dried prior to measurement of R_f values.

Consult: Hames (1981) or See and Jackowski (1989).

C. Protein Quantitation (Densitometry)

Quantitation of amounts of individual proteins in a mixture is often accomplished by scanning stained protein bands on a polyacrylamide gel with a spectrophotometer, and then integrating the peaks. Peak integration may be achieved electronically with some densitometers or manually by cutting the peaks from the chart paper and weighing. The absorbance of a protein band is proportional to the amount of protein only over a limited range of protein concentration. Proteins bind dyes with varying affinity, so comparison of peaks from different proteins should be based on standard curves for absolute quantitation when possible. Smith (1984) provides suggestions for the analysis of densitometric scans. Commercial gel scanning devices are available from Bio-Rad, Hoefer, Joyce-Loebl, and Pharmacia-LKB.

Protein quantitation may also involve elution of dye from stained gel bands. Ball (1986) and Wong et al. (1985) describe simple methods for removing Coomassie blue R-250 from polyacrylamide gel slices. It is important to note that absolute quantitation requires the generation of a standard curve with the protein of interest due to differences in dye binding by different proteins. The procedure of Ball (1986) is presented below:

- **Protocol**

1. Cut out protein band from the gel with a razor blade.

2. Place gel slice in a glass test tube, add 1ml of 3% SDS in 50% isopropanol, and cover with Parafilm.

3. Incubate in a 37°C water bath for 24 hours without agitation.

4. Remove liquid and determine absorbance at 595nm.

Consult: Ball (1986), Smith (1984), or Wong et al. (1985).

D. Eluting Protein Bands Following Electrophoresis

Extraction of protein from acrylamide gels can be accomplished by passive protein diffusion or electroelution. A simple procedure for passive protein elution (Bhown and Bennett, 1984) involves cutting the gel slice into small pieces with a razor, incubating for various times (15 minutes is a good period) in an appropriate buffer, followed by centrifugation, collection of the supernatant, and repeating the extraction a second time with more buffer. A better recovery yield can be expected with electroelution. A simple electroelution procedure termed reverse electrophoresis employing a tube gel is described by Otto and Snejdarkova (1981) and is adapted for slices from slab gels by Stralfors and Belfrage (1983). Electroelution devices are also available commercially from Bio-Rad and Pharmacia-LKB.

Reconstitution of enzyme activity following SDS-PAGE has been reported both *in situ* and after protein elution from gel slices (see also Chapter 6, Section III.B.). SDS removal sometimes requires replacement by a less denaturing detergent (nonionic or zwitterionic, see Chapter 1, Section IV.). Successful renaturation may depend on the concentrations of salt, glycerol, reducing agent, cofactor, or substrate in the renaturation buffer. A good discussion of enzyme renaturation following SDS-PAGE is found in Rothe and Maurer (1986, pp. 108-112).

Consult: Bhown and Bennett (1983), Otto and Snejdarkova (1981), Rothe and Maurer (1986), or Stralfors and Belfrage (1983).

V. Suppliers

• Equipment

1. Minigel apparatus:
   Bio-Rad Mini-Protean II Electrophoresis Cell
   Hoefer Scientific
   Pharmacia Midget System

2. Power supply (capacity 200V, 400mA):
   Bio-Rad Model 250/2.5 Power Supply
   Pharmacia EPS 500/400 or GPS 200/400 Power Supply

3. Boiling water bath or 100°C sand bath: Baxter

4. Eppendorf centrifuge:
   Baxter
   Fisher

5. Hamilton Syringes: Hamilton

6. Gel dryer:
   BioDesign, Inc.
   Bio-Rad Model 443 and 483
   Hoefer Drygel
   Pharmacia GSD-4 Gel Slab Dryer

7. High vacuum pump or water pump:
   Baxter
   Hoefer

8. Rocking or rotary shaker: Hoefer

• Reagents

1. Reagents for Gel Electrophoresis:

   Aldrich
   Bio-Rad
   Calbiochem
   Fluka
   Kodak
   Merck
   Pharmacia
   Sigma
   Whatman

2. Silver Staining Kits:

   Bio-Rad
   Pierce
   Polysciences

VI. References

SDS Polyacrylamide Gel Electrophoresis

Allen, R.C., C.A. Saravis, and H.R. Maurer. 1984. Gel Electrophoresis and Isoelectric Focusing of Proteins: Selected Techniques. 255 pages. Walter de Gruyter, Berlin.

Bio-Rad Bulletin 1089. 1984. Bio-Rad Silver Stain.

Bio-Rad Bulletin 1156. 1984. Acrylamide Polymerization - A Practical Approach.

Bio-Rad Mini Protean Slab Gel Instruction Manual.

Blackshear, P.J. 1984. Meth. Enzymol. 104: 237-255. Systems for Polyacrylamide Gel Electrophoresis.

Dion, A.S. and A.A. Pomenti. 1983. Anal. Biochem. 129: 490-496. Ammoniacal Silver Staining of Proteins: Mechanism of Glutaraldehyde Enhancement.

Giulian, G.G., R.L. Moss, and M. Greaser. 1983. Anal. Biochem. 129: 277-287. Improved Methodology for Analysis and Quantitation of Proteins on One-Dimensional Silver-Stained Slab Gels.

Gooderham, K. 1984. pp. 113-118 in Methods in Molecular Biology. Volume 1: Proteins. J.M. Walker, ed. 365 pages. Humana Press, Clifton, New Jersey.

Hames, B.D. pp. 1-91 in Hames, B.D and D. Rickwood, eds. 1981. Gel Electrophoresis of Proteins: A Practical Approach. 290 pages. IRL Press, Oxford and Washington, D.C.

Laemmli, U.K. 1970. Nature 227: 680-685. Cleavage of Structural Proteins during the Assembly of the Head of Bacteriophage T4.

Marshall, Thomas and Katherine M. Williams. 1984. Anal. Biochem. 139: 502-505. Artifacts Associated with 2-Mercaptoethanol upon High Resolution Two-Dimensional Electophoresis.

Merck Index, The. 1983. 10th Edition. M. Windholz, Ed. Merck & Co., Inc. Rahway, NJ.

Merril, C.R., D. Goldman, and M.L. Van Keuren. 1984. Meth. Enzymol. 104: 441-446. Gel Protein Stains: Silver Stain.

Ohsawa, K. and N. Ebata. 1983. Anal. Biochem. 135: 409-415. Silver Stain for Detecting 10-Femtogram Quantities of Protein after Polyacrylamide Gel Electrophoresis.

Perret, B.A., R. Felix, M. Furlan, and E.A. Beck. 1983. Anal. Biochem. 131: 46-50. Silver Staining of High Molecular Weight Proteins on Large-Pore Polyacrylamide Gels.

Pharmacia Publication. Polyacrylamide Gel Electrophoresis: Laboratory Techniques. 72 pages.

Rothe, G.M. and W.D. Maurer. 1986. pp. 37-140 in Gel Electrophoresis of Proteins. 407 pages. M.J. Dunn, ed. IOP Publishing Limited, Bristol, England.

Schleif, R.F. and P.C. Wensink. 1981. pp. 78-84. Practical Methods in Molecular Biology. 220 pages. New York, Springer-Verlag.

Scopes, R.K. 1982. pp. 163-171, 245-254. Protein Purification: Principles and Practice. 282 pages. Springer-Verlag, New York.

See, Y.P. and G. Jackowski. 1989. pp. 1-22 in Protein Structure: A Practical Approach. 355 pages. T.E. Creighton, ed. IRL Press, Oxford, England.

Sinclair, J. and D. Rickwood. pp. 189-218 in Hames, B.D. and D. Rickwood, eds. 1981. Gel Electrophoresis of Proteins: A Practical Approach. 290 pages. IRL Press, Oxford and Washington, D.C.

Smith, B.J. 1984. p. 141-146 in Methods in Molecular Biology. Volume 1: Proteins. J.M. Walker, ed. 365 pages. Humana Press, Clifton, New Jersey.

Smith, I., R. Cromie, and K. Stainsby. 1988. Anal. Biochem. 169: 370-371. Seeing Gel Wells Well.

Spanos, A. and U. Huebscher. 1983. Meth. Enzymol. 91: 263-277. Recovery of Functional Proteins in Sodium Dodecyl Sulfate Gels.

Weber, K., J.R. Pringle, and M. Osborn. 1971. Meth. Enzymol. 26: 3-27. Measurement of Molecular Weights by Electrophoresis on SDS-Acrylamide Gel.

Wedrychowski, A., R. Olinski, and L.S. Hnilica. 1986. Anal. Biochem. 159: 323-328. Modified Method of Silver Staining of Proteins in Polyacrylamide Gels.

Wilson, C.M. 1983. Meth. Enzymol. 91: 236-246. Staining of Proteins on Gels: Comparisons of Dyes and Procedures.

Wray, W., T. Boulikas, V.P. Wray, and R. Hancock. 1981. Anal. Biochem. 118: 197-203. Silver Staining of Proteins in Polyacrylamide Gels.

Gradient Gels

Hames, B.D. 1981. pp. 71-77 in Gel Electrophoresis of Proteins: A Practical Approach. Hames, B.D. and D. Rickwood, eds. 290 pages. IRL Press, Oxford and Washington D.C.

Walker, J.M. 1984. pp. 57-62 in Methods in Molecular Biology. Volume 1: Proteins. J.M. Walker, ed. 365 pages. Humana Press, Clifton, New Jersey.

SDS-Urea Gels

Schleif, R.F. and P.C. Wensink. 1981. pp. 84-87 in Practical Methods in Molecular Biology. 220 pages. Springer-Verlag, New York.

Swank, R.T. and K.D. Munkres. 1971. Anal. Biochem. 39: 462-477. Molecular Weight Analysis of Oligopeptides by Electrophoresis in Polyacrylamide Gel with Sodium Dodecyl Sulfate.

Detection of Radiolabeled Samples

Bonner, W.M. 1984. Meth. Enzymol. 104: 460-466. Fluorography for the Detection of Radioactivity in Gels.

Hames, B.D. 1981. pp. 49-59 in Gel Electrophoresis of Proteins: A Practical Approach. Hames, B.D. and D. Rickwood, eds. 290 pages. IRL Press, Oxford and Washington, D.C.

Harlow, E. and D. Lane. 1988. pp. 647-648 in Antibodies: A Laboratory Manual. 726 pages. Cold Spring Harbor Laboratory, Cold Spring Harbor, New York.

Roberts, P.L. 1985. Anal. Biochem. 147: 521-524. Comparison of Fluorographic Methods for Detecting Radioactivity in Polyacrylamide Gels or on Nitrocellulose Filters.

Waterborg, J.H. and H.R. Matthews. 1984. pp. 147-152 in Methods in Molecular Biology. Volume 1: Proteins. J.M. Walker, ed. 365 pages. Humana Press, Clifton, New Jersey.

Molecular Weight Determination

Hames, B.D. 1981. pp. 14-17 in Gel Electrophoresis of Proteins: A Practical Approach. Hames, B.D. and D. Rickwood, eds. 290 pages. IRL Press, Oxford and Washington, D.C.

See, Y.P. and G. Jackowski. 1989. pp. 10-18 in Protein Structure: A Practical Approach. T.E. Creighton, ed. 355 pages. IRL Press, Oxford.

Protein Quantitation (Densitometry)

Ball, E.H. 1986. Anal. Biochem. 155: 23-27. Quantitation of Proteins by Elution of Coomassie Brilliant Blue R from Stained Bands after Sodium Dodecyl Sulfate - Polyacrylamide Gel Electrophoresis.

Smith, B.J. 1984. pp. 119-125 in Methods in Molecular Biology. Volume 1: Proteins. J.M. Walker, ed. 365 pages. Humana Press, Clifton, New Jersey.

Wong, P., A. Barbeau, and A.D. Roses. 1985. Anal. Biochem. 150: 288-293. A Method to Quantitate Coomassie Blue-Stained Proteins in Cylindrical Polyacrylamide Gels.

Eluting Protein Bands Following Electrophoresis

Bhown, A.S. and J.C. Bennett. 1983. Meth. Enzymol. 91: 450-455. High-Sensitivity Sequence Analysis of Proteins Recovered from Sodium Dodecyl Sulfate Gels.

Otto, M. and M. Snejdarkova. 1981. Anal. Biochem. 111: 111-114. A Simple and Rapid Method for the Quantitative Isolation of Proteins from Polyacrylamide Gels.

Rothe, G.M. and W.D. Maurer. 1986. pp. 37-140 in Gel Electrophoresis of Proteins. M.J. Dunn, ed. 407 pages. IOP Publishing Ltd., Bristol, England.

Stralfors, P. and P. Belfrage. 1983. Anal. Biochem. 128: 7-10. Electrophoretic Elution of Proteins from Polyacrylamide Gel Slices.

Chapter 6

Gel Electrophoresis under Nondenaturing Conditions

I. Introduction

Nondenaturing gel electrophoresis, also called native gel electrophoresis, separates proteins based on their size and charge properties. While the acrylamide pore size serves to sieve molecules of different sizes, proteins which are more highly charged at the pH of the separating gel have a greater mobility. This method is capable of separating molecules which differ by a single unit charge. In addition, the conditions for nondenaturing gel electrophoresis minimize protein denaturation, in contrast to SDS polyacrylamide gel electrophoresis described in Chapter 5.

It is important to appreciate the effects of a few variables on nondenaturing gel electrophoresis.

Gel porosity: Acrylamide concentrations in the gel may be varied from approximately 5% to 15% and acrylamide:bis-acrylamide ratios may vary from 20:1 to 50:1 to achieve different sieving effects.

Charge: Most proteins are negatively charged at pH 8.8, which is the common pH used for nondenaturing gel electrophoresis. Alternatively, electrophoresis may be carried out at slightly acidic pH, in which case the anode and and cathode should be reversed. Note that in order to recover a protein with biological activity, it is necessary to work in a pH range which is not harmful to the protein of interest. For a practical discussion of pH effects on electrophoresis, see Hames (1981).

Ionic Strength: The ionic strength plays an important role in electrophoresis. If the ionic strength is too high there will be increased heat generation during electrophoresis but if the ionic strength is too low proteins may aggregate nonspecifically. Typically, ionic strengths in the 10-100mM range are utilized.

Temperature: All steps should be performed at 0 - 4°C to reduce loss of protein activity by denaturation and to minimize attack by proteolysis.

II. Discontinuous Nondenaturing Gel Electrophoresis

A. Introduction

Nondenaturing gel electrophoresis is commonly run with high pH buffers (pH 8.8). At this pH, most proteins are negatively charged and migrate toward the anode. The instructions are written for the Bio-Rad Mini-Protean apparatus, but all protocols should be easily adaptable to other systems.

B. Equipment

- Bio-Rad Mini-Protean Gel Electrophoresis Cell
- Power Supply
- Hamilton syringe
- Small container for staining and destaining the gel

Consult Chapter 5 for a list of suppliers.

C. Preparing the Gel

- **Reagents**

1. Acrylamide
2. Bis-acrylamide
3. Tris
4. Hydrochloric acid (HCl)
5. Ammonium persulfate
6. TEMED
7. Glycine
8. Glycerol
9. Bromophenol blue
10. Coomassie blue R-250 (for Coomassie blue staining)
11. Methanol
12. Glacial acetic acid

- **Working Solutions**

1. Solution A (Acrylamide Stock Solution), 100ml

   30% Acrylamide, 0.8% Bis-acrylamide

   Caution: Unpolymerized acrylamide is a skin irritant and a neurotoxin. Always handle with gloves.

   a. 30g acrylamide
   b. 0.8g bis-acrylamide

   Add distilled water to make 100ml and stir until completely dissolved. Work under hood and keep acrylamide solution covered with Parafilm until acrylamide powder is completely dissolved.

2. Solution B (4x Separating Buffer), 100ml

   1.5M Tris-Cl (pH 8.8)

   a. 18.2g Tris in 40ml H_2O
   b. add HCl to pH 8.8
   c. add H_2O to 100ml

3. Solution C (4x Stacking Buffer), 100ml

   0.5M Tris (pH 6.8)

   a. 6.0g Tris in 40ml H_2O
   b. add HCl to pH 6.8
   c. add H_2O to 100ml

4. 10% Ammonium Persulfate, 5ml

   a. 0.5g ammonium persulfate
   b. 5ml H_2O

5. Electrophoresis Buffer, 1 liter

a. 3.0g Tris	---25mM
b. 14.4g glycine	---192mM
c. H_2O to 1 liter (final pH should be 8.8)	

6. 5x Sample Buffer, 10ml

a. 3.1ml 1M Tris-Cl (pH 6.8)	---312.5mM
b. 5ml glycerol	---50%
c. 0.5ml 1% bromophenol blue	---0.05%
d. 1.4ml H_2O	

• Amounts of Working Solutions to Use

1. Volumes necessary for pouring gels of different thicknesses

(for two 6 x 8cm gels)

Gel Thickness	Separating	Stacking
0.5mm -	5.6ml	1.4ml
0.75mm -	8.4ml	2.1ml
1.0mm -	11.2ml	2.8ml
1.5mm -	16.8ml	4.2ml

Always prepare with a moderate excess.

2. Calculation for X% Separating Gel

Solution A	$^{x}/_{3}$ ml
Solution B	2.5ml
H_2O	$(7.5-^{x}/_{3})$ml
10% Ammonium Persulfate	50µl
TEMED	5µl (10µl if x<8%)
Total Volume	10ml

- **Pouring the Separating Gel**

Example of Separating Gel Preparation

Two 8% Separating Gels (6 x 8cm x 0.75mm), 10ml

4.8ml H_2O
2.7ml Solution A
2.5ml Solution B

50µl 10% Ammonium Persulfate
5µl TEMED

Do not prepare until following the numbered instructions below.

The instructions given below are very similar to those for denaturing gel electrophoresis in Chapter 5. Consult Chapter 5 for greater detail.

1. **Assemble gel sandwich** according to the manufacturer's instructions.

2. **Combine Solutions A and B and water** in a small Erlenmeyer flask or a test tube. Acrylamide (in Solution A) is a neurotoxin, so plastic gloves should be worn at all times.

3. **Add ammonium persulfate and TEMED, and mix** by swirling or inverting container gently. Work rapidly at this point because polymerization will be under way.

4. Carefully **introduce solution into gel sandwich** using a pipet (Fig. 6.1).

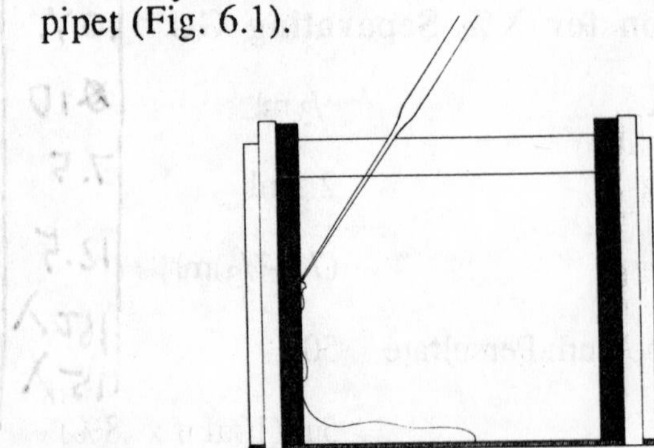

Figure 6.1. Introducing the separating gel solution into the gel sandwich, as presented in Fig. 5.2.

5. When the appropriate amount of separating gel solution has been added, gently **layer about 1cm of water** on top of the separating gel solution.

6. **Allow gel to polymerize** (30 - 60 minutes).

   When the gel has polymerized, a distinct interface will appear between the separating gel and the water, and the gel mold can be tilted to verify polymerization.

- **Comments on Separating Gels**

1. A low pH gel system for separation of basic proteins has been described by Reisfeld et al. (1962) with the following buffer composition:
   Separating gel: 0.06N KOH, 0.376M acetic acid, pH 4.3 (7.7%T, 2.67%C)
   Stacking gel: 0.06N KOH, 0.063M acetic acid, pH 6.8 (3.125%T, 25%C)
   Electrophoresis buffer: 0.14M β-alanine, 0.35M acetic acid, pH 4.5.

2. Remember to reverse the electrode polarity for low pH gels.

3. For the low pH gel system, methyl green should be used as the tracking dye in the sample buffer (at 0.002% final concentration).

4. Riboflavin may be used instead of ammonium persulfate for photopolymerization. Polymerization with riboflavin will leave fewer reactive compounds, but polymerization may not be as complete. A final riboflavin concentration of 0.0005% (from a 0.004% stock solution) is often utilized in the stacking gel, and polymerization is initiated by exposure to daylight or a white or blue fluorescent lamp.

5. If a reducing agent is required, 1mM DTT is recommended.

- **Pouring the Stacking Gel**

Example of Stacking Gel Preparation

Two 5% Stacking Gels (6 x 8cm x 0.75mm), 4ml
2.3ml H_2O
0.67ml Solution A
1.0ml Solution C

30µl 10% Ammonium Persulfate
5µl TEMED

Do not prepare until following instructions in section below.

1. **Pour off water** covering the separating gel.
2. **Combine Solutions A and C and water** in a small Erlenmeyer flask or a test tube.
3. **Add ammonium persulfate and TEMED, and mix** by gently swirling or inverting the container.
4. **Pipet stacking gel solution onto separating gel** until solution reaches top of front plate (Fig. 6.2).
5. **Carefully insert comb** into gel sandwich until bottom of teeth reach top of front plate (Figs. 6.3 and 6.4). Be sure no bubbles are trapped on ends of teeth.
6. **Allow stacking gel to polymerize** (about 30 minutes).
7. After stacking gel has polymerized, **remove comb** carefully (making sure not to tear the well ears).
8. Place gel into electrophoresis chamber. If using the Mini-Gel system, **attach both gels to electrode assembly before inserting into electrophoresis tank.**
9. **Add electrophoresis buffer to inner and outer reservoir**, making sure that both top and bottom of gel are immersed in buffer.

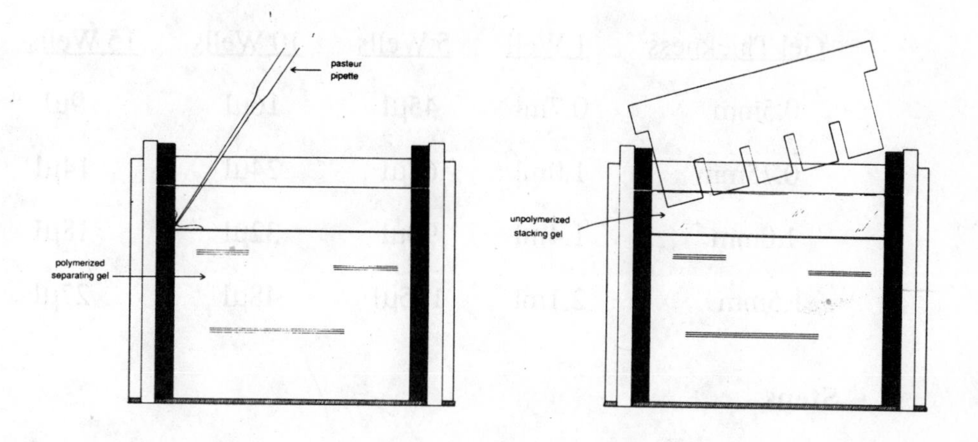

Figure 6.2. Introducing the stacking gel solution into the gel sandwich, as presented in Fig. 5.4.

Figure 6.3. Inserting the sample well comb into the stacking gel, as presented in Fig. 5.5.

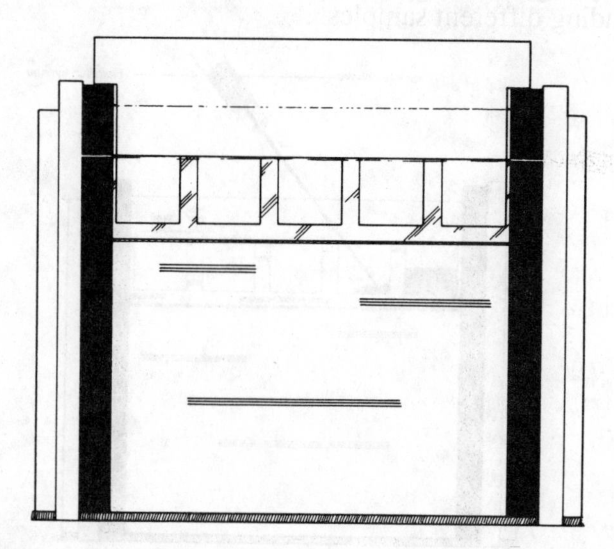

Figure 6.4. Stacking gel prior to polymerization, as presented in Fig. 5.6.

D. Sample Preparation

• **Capacity per Well** (Mini-Gel System)

Gel Thickness	1 Well	5 Wells	10 Wells	15 Wells
0.5mm	0.7ml	45μl	16μl	9μl
0.75mm	1.0ml	68μl	24μl	14μl
1.0mm	1.4ml	90μl	32μl	18μl
1.5mm	2.1ml	135μl	48μl	27μl

• **Steps**

1. **Combine protein sample and 5x Sample Buffer** (i.e., 20μl + 5μl) in an Eppendorf tube.
2. **Introduce sample solution into well** using a Hamilton syringe (Fig. 6.5).

Rinse syringe thoroughly with electrode buffer or water before loading different samples.

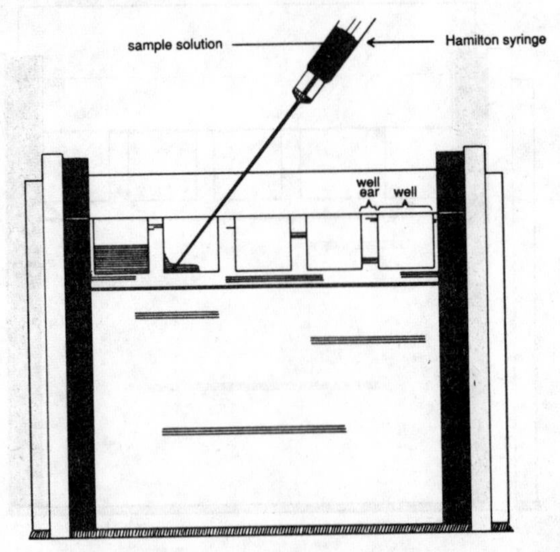

Figure 6.5. Introducing protein solution into sample well, as presented in Fig. 5.8.

• **Comments on Sample Preparation**

1. Typically, 1 - 5µg of protein is loaded per well, or up to 30µg of a complex protein mixture.

2. It is a good practice to fill unused gel lanes with blank sample buffer.

3. Sample may be centrifuged for 15 min at 10,000 x g prior to preparation for loading in order to remove insoluble material which may interfere with electrophoresis (Hames).

4. Samples with salt concentrations in excess of 0.1M may cause band distortion.

E. Running the Gel

- **Steps**

1. **Attach electrode plugs** to proper electrodes. Current should flow towards the anode for pH 8.8 gels.

2. **Turn on power supply** to 100 - 200V (constant current).

3. Electrophoresis should continue until the dye front migrates to 1cm from the bottom of the gel.

The high electrical current used in gel electrophoresis is very dangerous. **Never disconnect electrodes before first turning off the power source.** If using an electrophoresis apparatus which is not completely shielded from the environment, always leave a clearly visible sign warning that electrophoresis is in progress.

---

4. **Turn off power supply.**

5. **Remove electrode plugs** from electrodes.

6. **Remove gel plates** from electrode assembly.

7. Carefully remove a spacer, and, inserting the spacer in one corner between the plates, gently **pry apart the gel plates**. The gel will stick to one of the plates.

- **Comments**

1. Constant voltage will allow constant protein mobility during electrophoresis.

2. If the current is too high, excess heating may denature the protein, while current which is too low increases the time of electrophoresis and diffusion of bands. 100 - 200V is the recommended range for electrophoresis.

F. Staining the Gel

We describe Coomassie blue staining of nondenaturing gels (see Fig. 6.6). Coomassie blue staining can detect as little as 0.1µg of protein in a single band. Greater sensitivity is possible with the use of silver staining. For information about silver staining and more information about Coomassie blue staining, see Chapter 5, Section I.F.

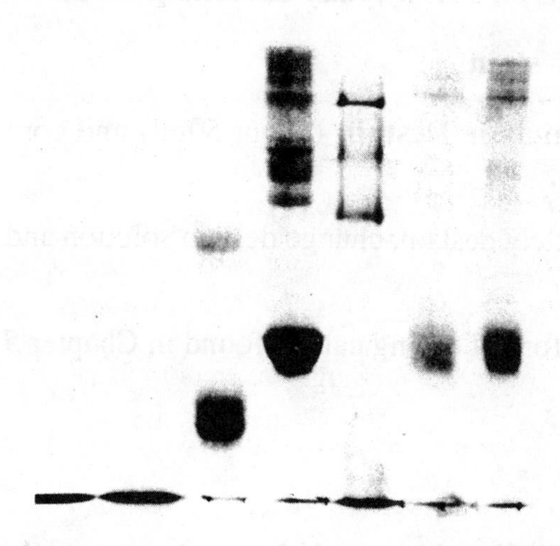

Figure 6.6. Nondenaturing gel stained with Coomassie blue. The gel concentration is 6% and was run with an assortment of protein markers from the Sigma nondenatured protein molecular weight marker kit.

- **Stock Solutions** (same as for SDS-PAGE)

1. <u>Coomassie Gel Stain</u>, 1 liter
   1.0g Coomassie blue R-250
   450ml methanol
   450ml H_2O
   100ml glacial acetic acid

2. <u>Coomassie Gel Destain</u>, 1 liter
   100ml methanol
   100ml glacial acetic acid
   800ml H_2O

• Staining Procedure

1. Wearing gloves, **pick up the gel and transfer to a small container** containing Coomassie Stain (20ml is sufficient).

2. **Agitate for 5-10 minutes** for 0.75mm gels or 10 - 20min for 1.5mm gels on a slow rotary or rocking shaker.

3. **Pour out stain.**

4. **Add Coomassie Destain** (about 50ml) and continue slow shaking.

5. To completely destain, change destain solution and agitate overnight.

Instructions for gel drying can be found in Chapter 5 (Section I.H.).

G. A Variant: Continuous Nondenaturing Gel Electrophoresis

Continuous gel electrophoresis is somewhat easier to perform than discontinuous gel electrophoresis since no stacking gel is involved. Instead, the gel sandwich is filled with separating gel solution only before the comb is inserted. However, the lack of stacking often results in thick, poorly resolved bands. It is important in continuous gel electrophoresis that the ionic strength of the protein buffer be 5 - 10 times less concentrated than the gel buffer in order to obtain the sharpest bands. Protein samples should fill no more than two or three millimeters of the well.

Buffers for continuous gel electrophoresis may be the same as described for discontinuous gel electrophoresis. Additional buffers are described by McLellan (1982).

III. Related Methods

A. Determining Protein Molecular Weight

Molecular weight determination is most commonly done by SDS-polyacrylamide gel electrophoresis where proteins are separated primarily by size. For molecular weight determination using nondenaturing gel electrophoresis, the protein sample must be run under a variety of acrylamide concentrations, often ranging from 4% to 12%. The accumulated information from these conditions serves to reduce the effect due to protein charge.

A simplified description of the data treatment follows (for a more detailed version, see Hedrick and Smith, 1968). Protein mobilities are calculated as the R_f value (distance of protein migration divided by distance of migration of the dye front). A semilogarithmic plot of the R_f relative to the acrylamide concentration (Fig. 6.7) should provide a line with a slope characteristic for a protein of a specific molecular weight. Proteins of known molecular weight should be electrophoresed under the same conditions, and the slopes generated from these experiments define a linear relationship with the molecular weight (Fig. 6.8). The molecular weight of the unknown protein of interest may be extrapolated from the data with the molecular weight standards.

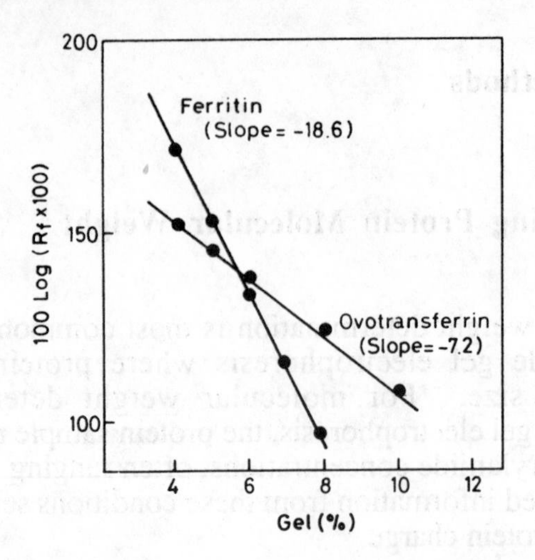

Figure 6.7. Graph of a logarithmic function of the relative mobility, 100 Log (R_f x 100), versus acrylamide concentration, i.e. gel (%). Modified from Hedrick and Smith (1968).

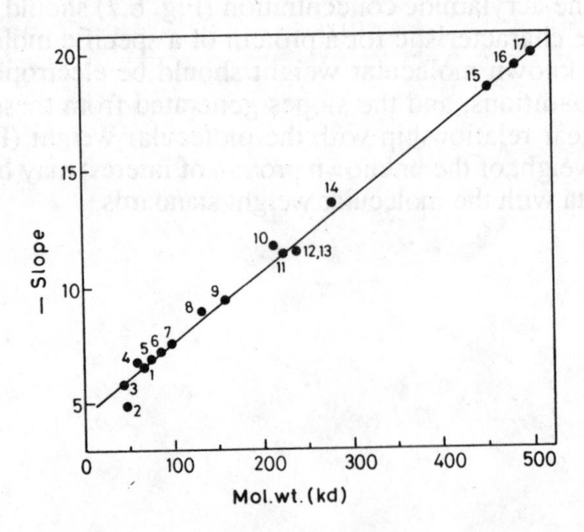

Figure 6.8. Graph of - slope of 100 log (R_f x 100) versus gel concentration (%) as a function of molecular weight. The slopes were obtained for the 17 proteins listed from graphs of the form presented in Fig. 6.7. The proteins included are 1. pepsin, 2. ovalbumin, 3. α-amylase, 4. albumin, 5. transferrin, 6. ovotransferrin, 7. hexokinase, 8. lactate dehydrogenase, 9. ketose-1-phosphate aldolase, 10. β-amylase, 11. nicotinamide deaminase, 12. α-urease, 13. catalase, 14. xanthine oxidase, 15. apoferritin, 16. urease, and 17. ribulose diphosphate carboxylase. Modified from Hedrick and Smith (1968).

B. Determining Enzyme Activity after Electrophoresis

Enzyme activity may be assayed after polyacrylamide gel electrophoresis either within the gel or following protein elution from the gel. It is not to be assumed that a protein will retain its activity following all the operations of gel electrophoresis; however, many proteins have been shown to be active after electrophoresis.

Activities have been determined within the polyacrylamide gel for a wide variety of enzyme classes. A good reference for experimental details is Gabriel (1971) which describes gel localization of enzymes including methods for dehydrogenases, transferases, hydrolases, lyases, and isomerases.

In order to remove a protein from the gel matrix, it is first helpful to determine the location of the protein in the gel. While staining with Coomassie blue or silver is common practice for protein visualization, these treatments may denature or modify the protein. An alternative procedure is fluorescence visualization of protein bands as reported by Leibowitz and Wang (1984).

As discussed in Chapter 5 (Section IV.D.), either passive diffusion or electroelution can be used to extract proteins from polyacrylamide gels. For passive diffusion the protein is extracted after cutting the gel in small pieces with a razor and eliminating the gel debris by centrifugation (Bhown and Bennet, 1983). However, the electroelution procedures, such as reverse electroelution (Otto and Snejdarkova, 1981), may provide better yields.

Rothe and Maurer (1986, pp. 55-56) provide a referenced table of over 40 proteins which were separated by polyacrylamide gel electrophoresis and identified by their enzymatic activity. Hubby and Lewontin (1966) describe enzyme assays for use in polyacrylamide gels for esterase, malic dehydrogenase, glucose-6-phosphate dehydrogenase, alkaline phosphatase, α-glycerophosphate dehydrogenase, and leucine aminopeptidase. Similar descriptions for the detection of α-amylase and acetaldehyde oxidase are found in Prakash et al. (1969).

IV. References

Bhown, A.S. and J.C. Bennett. 1983. Meth. Enzymol. 91: 450-455. High-Sensitivity Sequence Analysis of Proteins Recovered from Sodium Dodecyl Sulfate Gels.

Gabriel, O. 1971. Meth. Enzymol. 22: 578-604. Locating Enzymes on Gels.

Hames, B.D. 1981. pp. 23-64 in Gel Electrophoresis of Proteins: A Practical Approach. B.D. Hames and D. Rickwood, eds. 290 pages. IRL Press, London, England.

Hedrick, J.L. and A.J. Smith. 1968. Arch. Biochem. Biophys. 126: 155-164. Size and Charge Isomer Separation and Estimation of Molecular Weights of Proteins by Disc Gel Electrophoresis.

Hubby, J.L. and R.C. Lewontin. 1966. Genetics 54: 577-594. A Molecular Approach to the Study of Genic Heterozygosity in Natural Populations. I. The Number of Alleles at Different Loci in *Drosophila pseudoobscura.*

Leibowitz, M.J. and R.W. Young. 1984. Anal. Biochem. 137: 161-163. Visualization and Elution of Unstained Proteins from Polyacrylamide Gels.

McLellan, T. 1982. Anal. Biochem. 126: 94-99. Electrophoresis Buffers for Polyacrylamide Gels at Various pH.

Otto, M. and M. Snejdarkova. 1981. Anal. Biochem. 111: 111-114. A Simple and Rapid Method for the Quantitative Isolation of Proteins from Polyacrylamide Gels.

Prakash, S., R.C. Lewontin, and J.L. Hubby. 1969. Genetics 61: 841-858. A Molecular Approach to the Study of Genic Heterozygosity in Natural Populations. IV. Patterns of Genic Variation in Central, Marginal and Isolated Populations of *Drosophila pseudoobscura.*

Reisfeld, R.A., V.J. Lewis, and D.E. Williams. 1962. Nature 195: 281. Disk Electrophoresis of Basic Proteins and Peptides on Polyacrylamide Gels.

Rothe, G.M. and W.D. Maurer. 1986. pp. 37-140 in Gel Electrophoresis of Proteins. M.J. Dunn, ed. 407 pages. IOP Publishing Limited, Bristol, England.

Chapter 7

Isoelectric Focusing and Two Dimensional Gel Electrophoresis

I. Isoelectric Focusing (IEF)

A. Introduction

Isoelectric focusing gel electrophoresis is a technique that separates proteins according to their net charge or isoelectric point. Separation is accomplished by placing the protein in a pH gradient generated by an electric field. Under these conditions, the protein migrates until it reaches a position in the pH gradient at which its net charge, or isoelectric point, is zero. A powerful addition to isoelectric focusing was first demonstrated by O'Farrell (1975) when an isoelectric focusing gel was placed over an SDS-polyacrylamide gel and the focused proteins were separated in a second dimension according to their molecular weights. Two dimensional polyacrylamide gel electrophoresis has opened new possibilities for separating and studying complex protein mixtures (see Section II of this chapter).

For isoelectric focusing, protein bands are found to resolve better when a high voltage gradient is established across the gel. A high voltage gradient can be maintained only if efficient gel cooling is achieved. This requires efficient heat transfer between the gel and the liquid surrounding it. We have chosen to highlight the use of slab gels for isoelectrophoresis rather than more traditional tube gels to take advantage of the superior heat transfer capabilities of slab gels. In addition, the use of slab gels allows easy comparison of several protein samples.

Since isoelectric focusing is extremely sensitive to charge differences, reproducibility requires that a protein be handled with great care to avoid any modification of the protein's chemical composition or structure during sample preparation. In addition, interactions of proteins with lipids or with other proteins may cause charge modifications which will result in shifted isoelectric mobilities or streaking in the gel. Unless specific protein-protein interactions are being studied and if the protein must be maintained in a functional state, it is standard practice to carry out electrophoresis in a denaturing gel system with urea. Further improvements in resolution may be obtained with the use of nonionic detergents.

• Theory: Allen et al. (1984), pp.63-70.

• Time Required:

Individual Steps:

Pouring the Gel	90 minutes
Focusing	3 hours
2nd Dimension	
Equilibrating the Gel	30 minutes
Loading the Gel	15 minutes
Electrophoresis	45 minutes
(not necessary to fix afterwards)	
Fixing the Gel	2 hours - overnight
Coomassie Staining	30 minutes (for major bands)

Total Time: 7 hours including complete destaining

B. Equipment

- Bio-Rad Mini-Protean Gel System or other slab mini-gel apparatus
- Power supply (capacity 200V, 500mA)
- Hamilton syringe
- Small container for fixing and staining gel

C. Preparing Focusing Gel

- **Reagents**

1. Acrylamide
2. Bis-acrylamide
3. Ampholyte solutions
4. Urea, sequencing grade
5. Ammonium persulfate
6. TEMED
7. Triton X-100
8. 2-Mercaptoethanol
9. Bromophenol blue
10. Phosphoric acid
11. Sodium hydroxide (NaOH)
12. Potassium chloride (KCl)
13. Trichloroacetic acid (TCA)
14. Coomassie blue R-250
15. Methanol
16. Acetic Acid

• Stock Solutions

1. Solution A: 30% (w/v) acrylamide, 1% (w/v) bis-acrylamide (see Chapter 5)

2. 20% Triton X-100

3. 10% Trichloroacetic acid

4. 1% Trichloroacetic acid

5. 1% Bromophenol blue

6. Coomassie Gel Stain (see Chapter 5, Section I.F.)

7. Coomassie Gel Destain (see Chapter 5, Section I.G.)

• Ampholyte Selection

Ampholytes are amphoteric compounds which are provided as a mixture of molecules with closely spaced isoelectric points. They consist of oligo-amino acids and oligo-carboxylic acids in the 600-900 dalton molecular weight range (Righetti, 1989).

Giulian et al. (1984) have devised a table of ampholyte blends for use in the preparation of isoelectric focusing gels with different ranges of pH. It is reproduced below:

pH Range	Ampholyte pH Range	% in Final Gel Mixture
pH 3.5-10	pH 3.5-10	2.4%
pH 4-6	pH 3.5-10 pH 4-6	0.4% 2%
pH 6-9	pH 3.5-10 pH 6-8 pH 7-9	0.4% 1% 1%
pH 9-11	pH 3.5-10 pH 9-11	0.4% 2%

- **Pouring the Gel**

Characteristics: 5% T, 3.3% C (for definition of %T and %C, see Chapter 5, Section I.I.); adapted from Robertson et al. (1987).

Example of denaturing isoelectric focusing gel preparation: Two 8 x 7cm x 0.75mm minigels, 12ml

This example is for a gradient from pH 4-6. See Ampholyte Selection section above for recommended ampholyte mix to use for other pH ranges.

5.4ml H_2O
2.0ml Solution A
48μl ampholyte solution, pH 3.5-10
240μl ampholyte solution, pH 4-6
6.0g ultrapure urea

25μl 10% ammonium persulfate
20μl TEMED

Do not prepare until following numbered instructions below.

1. **Assemble gel plates** according to manufacturer's instructions.

2. **Combine urea, water, Solution A, and ampholyte solution** in a small Erlenmeyer flask. Wear gloves when working with acrylamide, which is a neurotoxin.

3. **Mix well** without shaking vigorously. Urea dissolves more rapidly if solution is warmed slightly.

4. **Add ammonium persulfate and TEMED, and mix gently**. Polymerization is under way at this point, so work rapidly.

5. **Pour acrylamide solution into assembled gel plates,** taking care to pour slowly along a spacer so no bubbles are trapped. Fill gel plates to the rim with acrylamide solution.

6. **Insert comb** so that teeth are entirely surrounded by gel (Fig. 7.1). Be careful not to trap air bubbles in the teeth of the comb.

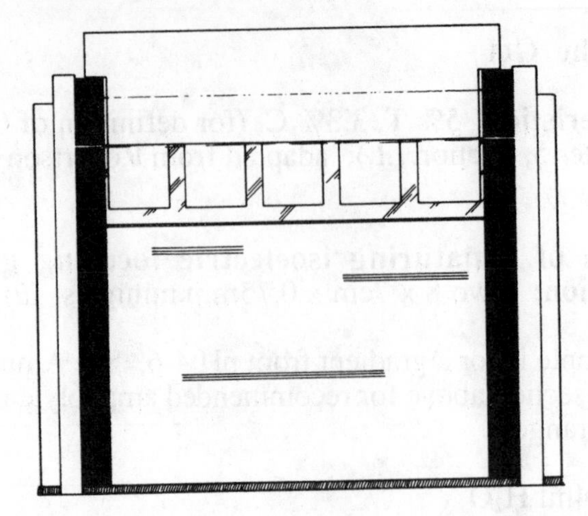

Figure 7.1. Stacking gel prior to polymerization, as presented in Fig. 5.6.

7. **Allow gels to polymerize** (about 1 hour).

8. After gels have polymerized, **remove comb** carefully.

9. **Attach gels to inner cooling core** and insert into electrophoresis tank. Before loading protein sample, it is useful to add anolyte (10mM phosphoric acid) to upper buffer chamber to verify that the gel assembly does not leak. To achieve a better seal, petroleum jelly may be applied to the sealing gasket as well as to the gel plates.

- **Comments**

1. Remove unpolymerized material at the bottom of the wells or it will polymerize during electrophoresis (Giulian et al., 1984).

2. Precast gels are commercially available, but only for horizontal slab electrophoresis systems (Pharmacia-LKB, Hoefer).

D. Sample Preparation and Loading (adapted from Robertson et al., 1987)

• **Capacity per Well** (Bio-Rad Mini-Gel System)

Gel Thickness	1 Well	5 Wells	10 Wells	15 Wells
0.5mm	0.7ml	45μl	16μl	9μl
0.75mm	1.0ml	68μl	24μl	14μl
1.0mm	1.4ml	90μl	32μl	18μl
1.5mm	2.1ml	135μl	48μl	27μl

• **Denaturing Gel Loading Buffer (2x)**, for pH gradient 4-6, 5ml

2.4g urea (8M)
20μl ampholyte solution, pH 3.5-10
100μl ampholyte solution, pH 4-6
500μl 20% Triton X-100 (2%)
50μl 2-mercaptoethanol (1%)
1.7ml distilled water
200μl 1% Bromophenol blue

Can be stored frozen at -20°C in 0.5ml aliquots (Pollard, 1984)

• **Steps**

1. Mix protein sample with an equal volume of 2x Loading Buffer. Before applying sample to gel, spin 5 minutes at 10,000 x g (in Eppendorf centrifuge) to remove aggregated protein which will cause streaking (Sinclair and Rickwood, 1981).

2. Apply the protein sample into the bottom of the well with a Hamilton syringe.

• **Comments**

1. For Coomassie staining, 10 - 30μg of protein from a crude mixture or 5 - 10μg of a single protein species per lane are reasonable loading estimates.

2. If samples do not go into solution readily, it is possible to sonicate them, making sure to keep the temperature below 30°C.

E. Running Isoelectric Focusing

• **Electrophoresis Solutions** (O'Farrell, 1975)

1. Add catholyte (20mM sodium hydroxide) to upper buffer chamber.

2. Add anolyte (10mM phosphoric acid) to lower buffer chamber.

• **Comments**

1. Anolyte should be made fresh from a 1M phosphoric acid stock solution and catholyte should be made fresh from 1M sodium hydroxide stock solution (Pollard, 1984).

2. Work at room temperature to prevent precipitation of urea.

• **Focusing Conditions** (from Robertson et al., 1987)

1. Attach electrodes.

2. Run for 30 minutes at 150V (constant voltage).

3. Then set at 200V for 2.5 hours (constant voltage). Current will be about 10mA at the start and will decrease during focusing. Inner chamber will heat to 40 - 50°C during the course of electrophoresis.

F. Post-Focusing Procedures

• Determining pH Gradient

1. Cut a strip of gel into 0.5cm or 1cm slices.
2. Suspend each slice in 1ml 10mM KCl for about 30 minutes.
3. Read pH of KCl solutions.

• Fixing the Gel (from Robertson et al., 1987)

1. Place gels in 10% trichloroacetic acid (TCA) for 10 minutes.
2. Replace with 1% TCA and soak for at least 2 hours (to remove ampholytes). Soaking overnight is best for reducing Coomassie staining of ampholytes.

• Staining the Gel

1. Stain gel for 10 minutes in Coomassie Gel Stain (see Chapter 5), rocking gently.

2. Remove Coomassie Gel Stain and replace with Coomassie Gel Destain (see Chapter 5). Replace Destain several times, allowing destaining to continue overnight.

3. Gel may be dried as described in Chapter 5. A typical gel is presented in Fig. 7.2.

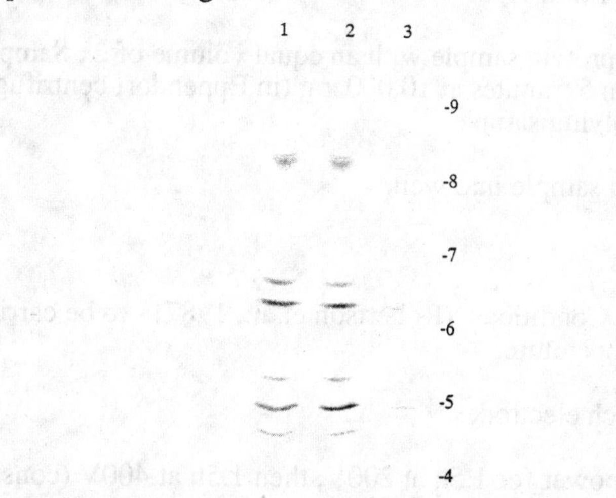

Figure 7.2. Isoelectric focusing gel published by Robertson et al., (1987), kindly provided by H.K. Dannelly.

G. Modifications for a Native Isoelectric Focusing Gel

To run a native isoelectric focusing gel, the following protocol modifications must be made:

• Pouring the Gel (pH 4-6)

Native Isoelectric Focusing Gel
5% T, 3.3%C (Two 8 x 7cm x 0.75mm minigels), 12ml
9.7ml H_2O
2ml Solution A
48µl ampholyte solution pH 3.5-10
240µl ampholyte solution pH 4-6

50µl 10% ammonium persulfate
20µl TEMED

• Sample Preparation and Application

Native Gel Sample Buffer (2x), 5ml
3ml glycerol
200µl ampholytes (same proportions as for gel)
1.8ml H_2O

1. Mix protein sample with an equal volume of 2x Sample Buffer. Spin 5 minutes at 10,000 x g (in Eppendorf centrifuge) before applying sample.

2. Load sample into well.

• Focusing Conditions (Robertson et al., 1987) - to be carried out at room temperature.

1. Attach electrodes.

2. Set power for 1.5h at 200V, then 1.5h at 400V (constant voltage).

H. Discussion

- **Troubleshooting** (adapted from Allen et al., 1984)

1. **Incomplete focusing** is evidenced by fuzzy bands. This may be due to problems in the electrophoresis or to large proteins which have restricted mobility in the gel. If focusing is carried out for too short or too long a period, band resolution is decreased. Increasing the voltage gradient incrementally towards the end of the run sharpens bands. High molecular weight proteins may focus better in more porous agarose gels (see Comments).

2. **Skewed bands** are usually due to faults in the pH gradient. Verify that the electrodes are clean and make good contact with the gel. Also, be aware of aberrant effects at the edge of the gel.

3. **Protein band streaking** is a recurrent problem in isoelectric focusing. There are a number of possible causes:

    a. Protein aggregation or precipitation, especially near the protein's isoelectric point, or sample overloading. 8M urea usually counteracts a tendency to aggregate. Detergents such as Triton X-100 and Nonidet P-40 are useful especially to solve aggregation problems with membrane proteins (see Comments). Be sure to centrifuge samples before loading gel to remove particulates.

    b. Presence of nucleic acids in the sample. Various procedures to remove nucleic acids, including acid extraction, salt precipitation, and nuclease treatment, are cited in Sinclair and Rickwood (1981, p.194).

    c. Protein modification. When the urea is not ultra-pure, isocyanate impurities may cause protein carbamoylation. Prerunning the gel can remove isocyanate. Other modifications include oxidation of Cys residues or deamination of Asn or Gln when the protein sample is improperly handled and stored.

4. **Wavy bands** are often due to a high salt content of the samples. Sometimes wavy bands are attributed to impurities in the ampholyte or electrolyte solutions, or to dirty electrodes.

5. **Uneven pH gradient** may result from electrode contact not being parallel to the gel, impurities within the gel, or ampholyte concentrations which are too low. If the pH gradient in the alkaline portion of the gel is lost, this is likely due to cathodic drift. Supplement with pH 9-11 ampholytes or run a nonequilibrium pH gradient electrophoresis gel (NEPHGE, see Comments).

6. **High background staining** is probably due to ampholytes remaining in the gel after fixation. Increase the time of fixing with 1% TCA.

7. **Missing or faint bands** are likely to be low molecular weight proteins (<10 kd) or proteins which have not been denatured during fixation. Increase TCA concentration or fix with glutaraldehyde (see Chapter 5).

8. **Overlapping spots** may occur in complex protein mixtures. Changing the pH range of the isoelectric focusing gel may solve this problem. Further steps of protein purification or immunoprecipitation are recommended to remove an overlapping spot.

- **Comments**

1. To establish the position of the dye front after isoelectric focusing, mark the dye front with a fine gauge wire (0.1mm diameter). The wire may be dipped in India ink for increased visibility.

2. Different brands of ampholytes have slightly different properties, and gels run with ampholytes from different sources will have differing patterns of protein separation. For the best reproducibility, do not change ampholyte brands.

3. In general, shallower pH gradients will lead to better resolution of protein bands. However, shallower pH gradients require a longer focusing time. A good compromise is a pH range of 2 (der Lan and Chrambach, 1981).

4. Because of the limitations of power supplies and gel cooling, the maximum recommended length for gels is 8 - 10cm. It is better to run several gels with pH gradients than a single long gel (Allen et al., 1984).

5. Cathodic drift becomes a problem when gels are subjected to electrophoresis for very long periods (>3000 volt-hours) due to lability of the ampholytes. Cathodic drift results in the partial collapse of the pH gradient, particularly at pH values above 8. To overcome this problem, O'Farrell et al. (1977) developed a technique in which isoelectric focusing gels were run for shorter periods (1600 volt-hours). This method permits focusing of proteins at higher pH ranges and is called nonequilibrium pH gradient electrophoresis (NEPHGE). However, it is not possible to determine a protein's isoelectric point using NEPHGE. See O'Farrell et al. (1977) or Phillips (1988) for instructions.

6. Urea solubilizes proteins and eliminates protein-protein and protein-lipid interactions. This allows separation to occur more rapidly and improves resolution. Urea also inhibits cathodic drift. It should be noted that the isoelectric point of a denatured protein may differ from that of the native protein.

7. If a protein solution contains SDS, add urea to the sample to a final concentration of 8M. At high concentrations of urea, SDS interacts minimally with proteins.

8. 2% Nonidet P-40 or 2% Triton X-100 (O'Farrell et al., 1977; Giulian et al., 1984) are often added to denaturing solutions to help keep proteins (especially membrane proteins) soluble. Some authors recommend zwitterionic detergents such as CHAPS and Zwittergent 3-14.

9. Ultra-thin gels (50 - 500μm thick) provide several advantages over standard slab gel isoelectric focusing. The thinness of the gel allows increased speed and resolution because of higher field strengths and improved cooling. Protocols may be found in Allen et al. (1984), Giulian et al. (1984), and Radola (1983).

10. Ready-made Immobiline isoelectric focusing gels (Pharmacia-LKB), in which the ampholytes are covalently bonded to the acrylamide, are available for high-resolution separation in the 4-7 pH range.

11. High molecular weight proteins (>750 kd) often have aberrant mobility in polyacrylamide isoelectric focusing gels. Agarose or agarose-acrylamide gels have provided satisfactory alternatives. For protocol descriptions, see Allen et al. (1984) and Pino and Hart (1984).

12. Coomassie staining may be improved if the TCA fixation step is followed by a 10 - 30 minute rinse of the gel in a solution of 0.25% SDS in ethanol:acetic acid:water (33:10:57). SDS binds ampholytes, leading to improved removal of the ampholytes from the gel (Giulian et al., 1984).

II. Two Dimensional Gel Electrophoresis

A. Introduction

The second dimension of two dimensional gel electrophoresis simply consists of SDS-polyacrylamide gel electrophoresis as described in Chapter 5. A strip of gel or a tube gel from isoelectric focusing (the first dimension) is fitted over an SDS-polyacrylamide gel and the proteins are separated according to molecular weight by electrophoresis. Pre-equilibration of the isoelectric focusing gel in SDS is necessary prior to running the second dimension.

B. Equipment

- Bio-Rad Mini-Protean Gel Electrophoresis Cell
- Power supply (capacity 200V, 500mA)
- Container for staining and destaining the gel

C. Protocols

- Reagents

  Glycerol
  2-Mercaptoethanol
  Sodium dodecyl sulfate (SDS)
  Tris
  For gel electrophoresis reagents, consult Chapter 5, Section I.C.

- Working Solutions

  1. 20% SDS
  2. 1M Tris-HCl (pH 6.8)

• Steps

1. If the second dimension gel is to be run immediately following isoelectric focusing, pour a 1.0 or 1.5mm SDS-polyacrylamide gel at the same time as pouring the isoelectric focusing gel. Instead of placing a comb into the stacking gel, leave about 0.5cm above the stacking gel and overlay very carefully with water. The isoelectric focusing gel may be stored frozen and run in the second dimension later (see below).

2. Run isoelectric focusing gel as described in the first part of this chapter.

3. After focusing, cut a strip of gel 0.5cm wide and place in a small container or a boat made of parafilm.

4. Equilibration

Equilibration Buffer, 100ml

5ml 2-Mercaptoethanol	(5%)
6.25ml 1M Tris-HCl (pH 6.8)	(62.5mM)
11.5ml 20% SDS	(2.3%)
10ml Glycerol	(10%)
H_2O to 100ml	

a. Add equilibration buffer to IEF gel.
b. Incubate for 15 - 30 minutes.
c. Gel can be immediately loaded onto an SDS-polyacrylamide gel or frozen and stored at -80°C.

5. Electrophoresis

a. Remove water overlay from stacking gel by aspiration.
b. Overlay the stacking gel with electrode buffer (see Chapter 5 for preparation details).
c. Wearing gloves and being careful not to distort the gel, place the isoelectric focusing gel strip on the stacking gel. It is a good idea to lower the gel strip gently onto the stacking gel from one side with a spatula or syringe to keep from trapping air bubbles. When loading gel, note which side is acidic and which is basic from isoelectric focusing.
d. Proceed with electrophoresis and staining as described in Chapter 5 (Sections I. F., G., and H.). A typical two-dimensional gel is presented in Fig. 7.3.

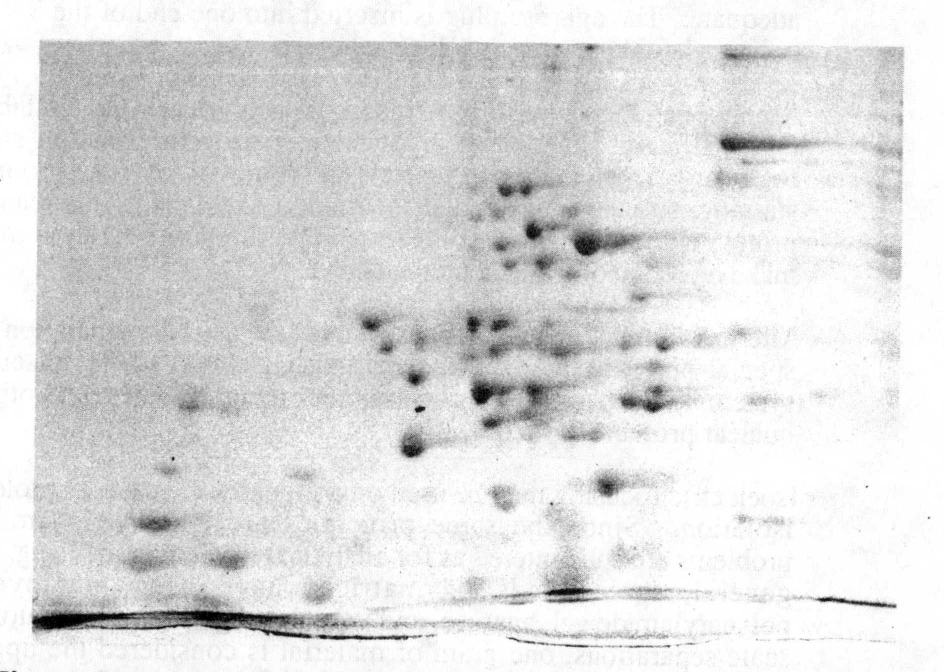

Figure 7.3. Two-dimensional gel of a crude extract of *Escherichia coli* stained with Coomassie blue. Separation in the horizontal dimension was achieved by isoelectric focusing in the pH range 4-7 in the presence of 8M urea, followed by separation in the vertical dimesion by SDS-polyacrylamide gel electrophoresis. From Robertson et al. (1987), kindly provided by H.K. Dannelly.

D. Discussion

- A stacking gel sometimes causes band elongation in the direction of isoelectric focusing due to lateral band spreading (Strahler et al., 1989). If this is a problem, the stacking gel may be omitted.

- Streaking in second dimension (SDS-PAGE) may be due to poor equilibration of the first dimension gel or insufficient buffering capacity of the Tris in the second dimension. 2-Mercaptoethanol may cause streaks with silver staining (see Chapter 5).

- To run molecular weight markers in the second dimension, embed the markers in a 1% agarose plug with 0.02% bromophenol blue. The plug should be poured in a tube of the right diameter, and then the solidified gel is cut into 0.5 - 1.0cm sections which can be stored frozen. 5 - 10μg of molecular weight proteins per slice are adequate. The agarose plug is inserted into one end of the SDS-polyacrylamide gel prior to electrophoresis.

- Ampholytes run as small proteins in the gel, are acid precipitable and will stain. Elution of ampholytes during the fixation step following focusing is essential for removal of background staining; however, extended incubation at this step may cause protein elution from the gel. Protein elution during fixation may make reproducibility difficult to achieve.

- Alternative two-dimensional gel systems are used for separation of special protein classes. Sinclair and Rickwood (1981) discuss systems for separation of ribosomal proteins, histones, and other nuclear proteins (pp. 209-217).

- Isoelectric focusing may be used on a preparative scale for protein isolation. Since the same principles are employed, similar problems are encountered as for analytical isoelectric focusing. In general, three kinds of matrices have been employed: polyacrylamide gel, agarose, and granulated gels. For laboratory-scale separations, one gram of material is considered the upper limit for separation, and this amount may be separated in under an hour with certain systems. For a good description of the method, see Radola, 1984. For protocols, consult Radola (1984), Allen et al. (1984), or der Lan and Chrambach (1981).

III. Suppliers

Electrophoresis Cell

- BioRad Mini-Protean II Electrophoresis Cell
- Hoefer Mighty Small II System
- Pharmacia Midget System

- Commercial apparatus for ultra-thin and preparative isoelectric focusing are available from BioRad, Hoefer, and Pharmacia-LKB.

Power Supply

- BioRad Model 250/2.5 Power Supply
- Hoefer PS500XT or PS500X Power Supply
- Pharmacia EPS 500/400 or GPS 200/400 Power Supply

Hamilton Syringes: Hamilton

Chemicals for gel electrophoresis

- Aldrich, BDH, BioRad, Calbiochem, Fluka, Kodak, Merck, Pharmacia, Sigma, Whatman

Ampholyte solutions

- BioRad: BioLytes
- Pharmacia-LKB: Ampholines, Immobilines, Pharmalytes
- Serva: Servalyts

IV. References

Allen, R.C., C.A. Saravis, and H.R. Maurer. 1984. pp. 63-147, Isoelectric Focusing and pp. 148-180, Multiparameter Techniques. In Gel Electrophoresis and Isoelectric Focusing of Proteins: Selected Techniques. 255 pages. Walter de Gruyter, Berlin.

Giulian, G.G., R.L. Moss, and M. Greaser. 1984. Anal. Biochem. 142: 421-436. Analytical Isoelectric Focusing Using a High-Voltage Vertical Slab Polyacrylamide Gel System.

der Lan, B. and A. Chrambach. 1981. pp. 157-188 in Gel Electrophoresis of Proteins: A Practical Approach. Hames, B.D. and D. Rickwood, eds. 290 pages. IRL Press, Oxford and Washington, D.C.

O'Farrell, P.H. 1975. J. Biol. Chem. 250: 4007-4021. High Resolution Two-Dimensional Elecrophoresis of Proteins.

O'Farrell, P.Z., H.M. Goodman, and P.H. O'Farrell. 1977. Cell 12: 1133-1142. High Resolution Two-Dimensional Electrophoresis of Basic as Well as Acidic Proteins.

Phillips, T.A. 1988. DNA and Protein Engineering Techniques 1: 5-9. Two-Dimensional Polyacrylamide Gel Electrophoresis of Proteins. Alan R. Liss, New York.

Pino, R.M. and T.K. Hart. 1984. Anal. Biochem. 139: 77-81. Isoelectric Focusing in Polyacrylamide-Agarose.

Pollard, J.W. 1984. pp. 81-96. Two-Dimensional Polyacrylamide Gel Electrophoresis of Proteins. pp. 81-96 in Methods in Molecular Biology, Vol. 1, Proteins, J.M. Walker, ed. 365 pages. Humana Press, Clifton, New Jersey.

Radola, B.J. 1983. pp. 101-118. Ultra-Thin-Layer Isoelectric Focusing. In Electrophoretic Techniques, C.F. Simpson and M. Whittaker, eds. 280 pages. Academic Press, London.

Radola, B.J. 1984. Meth. Enzymol. 104: 256-274. High-Resolution Preparative Isoelectric Focusing.

Righetti, P.G. 1989. pp. 23-63 in Protein Structure: A Practical Approach. T.E. Creighton, ed. 355 pages. IRL Press, Oxford.

Robertson, E.F., H.K. Dannelly, P.J. Malloy, and H.C. Reeves. 1987. Anal Biochem. 167: 290-294. Rapid Isoelectric Focusing in a Vertical Polyacrylamide Minigel System.

Sinclair, J. and D. Rickwood. 1981. Two-Dimensional Gel Electrophoresis. pp. 189-218 in Gel Electrophoresis of Proteins: A Practical Approach, B.D. Hames and D. Rickwood, eds. 290 pages. IRL Press, Oxford and Washington, D.C.

Strahler, J.R., R. Kuick, and S.M. Hanash. pp. 65-92 in Protein Structure: A Practical Approach. T.E. Creighton, ed. 355 pages. IRL Press, Oxford.

Chapter 8

Immunoblotting

I. Introduction

This chapter describes immunochemical techniques, termed immunoblotting or Western blotting, which are used to detect a protein immobilized on a matrix (Towbin et al., 1979). Before employing this procedure, it is necessary to have a monoclonal or polyclonal antibody capable of recognizing the protein of interest and a solution containing the protein, either a crude extract or a more purified preparation. Immunoblotting is an extremely powerful technique for identifying a single protein (or epitope) in a complex mixture following separation based on its molecular weight (SDS-PAGE, Chapter 5), size and charge (nondenaturing gel electrophoresis, Chapter 6) or isoelectric point (isoelectic focusing, Chapter 7). In addition, immunoblotting combined with immunoprecipitation permits the quantitative analysis of minor antigens. A number of other applications for immunoblots are listed below. The immobilization of proteins on a membrane matrix is preferred to working directly with the gel because the proteins are more accessible, membranes are easier to handle than gels, smaller amounts of reagents are needed, and processing times are shorter (Gershoni and Palade, 1982).

Limits of Detection: 10 pg (picogram = 10^{-12}g) with horseradish peroxidase or alkaline phosphatase labeling or 1 pg with immunogold or ^{125}I labeling.

Theory: Gershoni and Palade, 1983

Description of Immunoblotting

Immunoblotting can be divided into two steps: transfer of the protein from the gel to the matrix and decoration of the epitope with the specific antibody.

Protein transfer is most commonly accomplished by electrophoresis. The two common electrophoretic methods are:

1. Semi-dry blotting, in which the gel and immobilizing matrix are sandwiched between buffer-wetted filter papers through which a current is applied for 10 - 30 minutes.
2. Wet (tank) blotting, in which the gel-matrix sandwich is submerged in transfer buffer for electrotransfer, which may take as little as 45 minutes or may be allowed to continue overnight.

We present only wet blotting here, since it permits greater flexibility without being significantly more expensive in time or materials. The Bio-Rad Mini Trans-Blot transfer cell is used for describing the electrotransfer, but the conditions should be easily adaptable to other electroblotting devices.

Following transfer, detection of the epitope proceeds in two or three steps. First, the nitrocellulose membrane is incubated with the primary antibody for several hours or overnight. Next, a second antibody (Protein A may also be used), which recognizes an epitope on the first antibody is incubated with the nitrocellulose. Typically, the second antibody (usually goat antibodies raised against rabbit immunoglobulins for rabbit-generated first antibodies) is purchased already conjugated to a labeling agent such as the enzyme horseradish peroxidase. This marker is then visualized by a colorimetric reaction catalyzed by the enzyme which yields a colored product that remains fixed to the nitrocellulose membrane. Other detection systems including alkaline phosphatase or immunogold conjugates and ^{125}I labels are also described below.

Blotting Membranes

Two kinds of membranes are most commonly used for transferring proteins from gels: nitrocellulose and nylon. For most applications, nitrocellulose appears to be the membrane of choice and this chapter will describe this support only for protein immobilization. Nitrocellulose is preferred because it is relatively inexpensive and blocking nitrocellulose from nonspecific antibody binding is fast and simple.

Nylon may be more useful if **1.** a higher protein binding capacity is required (480μg/cm^2 vs. 80μg/cm^2 for nitrocellulose; Gershoni and Palade, 1982); **2.** the protein to be studied binds weakly to nitrocellulose (especially high molecular weight or acidic proteins); or **3.** greater resistance to mechanical stress is desired. The use of nylon has been limited because it is more expensive than nitrocellulose, blocking is cumbersome, and staining for total protein with anionic dyes is not possible.

Other Applications Involving Immobilized Proteins (see Discussion)

1. Epitope mapping
2. Structural domain analysis
3. Dot blotting
4. Renaturing proteins for functional assay
5. Ligand binding
6. Improved autoradiography
7. Antibody purification
8. Cutting out protein bands for antibody production
9. Protein identification: amino acid analysis and protein sequencing

II. Performing an Immunoblot

A. Equipment

- Electroblotting Apparatus
  We describe the Bio-Rad Mini Trans-Blot Electrophoretic Transfer Cell. Other commercially available or homemade apparatus can also be used.

- Power supply with capacity of 200V, 0.6A

- Whatman 3MM paper

- Nitrocellulose paper, 0.2 or 0.45μm pore size
  Nitrocellulose should be stored in a cool dark place and gloves should be worn when handling the membrane to prevent protein transfer.

- Seal-A-Meal apparatus

- Small plastic or glass container for gel incubations (15 x 10 x 2cm)

- Rocker or rotary shaker

- Shallow tray used in preparing gel for transfer (30 x 15 x 3cm)

- Magnetic stir plate

B. Reagents

- **Transfer Buffer:**
  1. Tris
  2. Glycine

- **Tris Buffered Saline (TBS):**
  1. Tris, 2M solution at pH 7.5, for instructions see Chapter 5 section I.C.
  2. Sodium chloride (NaCl), 4M solution

- **Blocking Solution**: Bovine serum albumin (BSA)

- **Horseradish Peroxidase Developing Reagent:**
  1. Chloronaphthol
  2. Methanol
  3. Hydrogen peroxide (H_2O_2)

- **Alkaline Phosphatase Developing Reagent:**
  1. $MgCl_2$
  2. 5-Bromo-4-chloro-3-indolyl phosphate (BCIP)
  3. Dimethylformamide (DMF)
  4. Veronal acetate buffer
  5. *p*-Nitro blue tetrazolium chloride (NBT)

- **Immunogold Incubation Buffer**: Tween 20

- **Amido Black Total Protein Stain:**
  1. Amido black 10B
  2. Isopropanol
  3. Acetic acid

- **India Ink Total Protein Stain:**
  1. Tween 20
  2. India Ink

- **Immunoblot Erasing Buffer:**
  1. Powdered nonfat milk
  2. Sodium dodecyl sulfate (SDS)
  3. Tris
  4. 2-Mercaptoethanol

C. Protocols

• Transfer

Time required: minimum of 90 minutes, or overnight

This protocol begins once a polyacrylamide gel (SDS, native or IEF) has already been run to separate the protein of interest from other proteins in the sample. It is recommended to run two such gels, staining one directly and using the other for immunoblotting. Alternatively, it is possible to stain the nitrocellulose membrane for total protein following electroblotting (see Section E).

Transfer Buffer, 1 liter

1.93g Tris	15.6mM
9g Glycine	120mM

Add distilled water to 1 liter, pH should be between 8.1 and 8.4 without adjustment.

Can also be made up as a 20x stock solution.

Preparations:

- Make up 1 liter transfer buffer and chill to 4°C.
- Put distilled water in cooling unit and freeze overnight.

1. **Prepare transfer cell and nitrocellulose:**
   a. Soak nitrocellulose sheet (6cm x 8cm) in transfer buffer (30 - 40ml) for 15 - 20 minutes.
   b. Rinse buffer chamber with distilled water.
   c. Insert Trans-Blot electrode insert and small stir bar into buffer chamber, and add transfer buffer until half full (about 400ml).
   d. Insert frozen cooling unit.

2. **Arrange polyacrylamide gel-membrane sandwich** (Fig. 8.1 and 8.2):
   a. In a shallow tray, open the transfer cassette and place a wetted sheet of Whatman 3MM paper (8 x 10cm) on a well-soaked fiber pad on the gray panel of the transfer cassette. Transfer buffer should be used for soaking.
   b. Carefully place the gel on the wet filter paper and arrange well ears and gel so that all air bubbles are removed. Use transfer buffer for lubrication and be sure to wear gloves (Fig. 8.3).
   c. After wetting gel, carefully lay a wetted sheet of nitrocellulose on top, beginning from one side so air bubbles are removed to the edge of the gel (Fig. 8.4). Be sure to wear gloves or use tweezers when handling membranes.
   d. Place a wetted sheet of 3MM paper over the nitrocellulose and roll a small test tube or pipette like a rolling pin over the sandwich to remove any air bubbles (Fig. 8.5).
   e. Cover with the second well-soaked fiber pad, close the transfer cassette and slide it into the electrode insert in the buffer tank, keeping the gray panel of the cassette on the same side as the gray panel of the electrode assembly (Fig. 8.6).
   f. Fill the buffer tank with transfer buffer.

3. **Electrotransfer**:
   a. Place entire Trans-Blot apparatus on a magnetic stir plate and begin stirring.
   b. Attach the electrodes.
   c. Set the power supply to 100V (constant voltage) and transfer for 1h. For chilled transfer buffer, initial current will be 0.2A and final current will be 0.4A.

- Overnight transfers should be run at 30V (Bio-Rad Trans-Blot Instruction Manual).

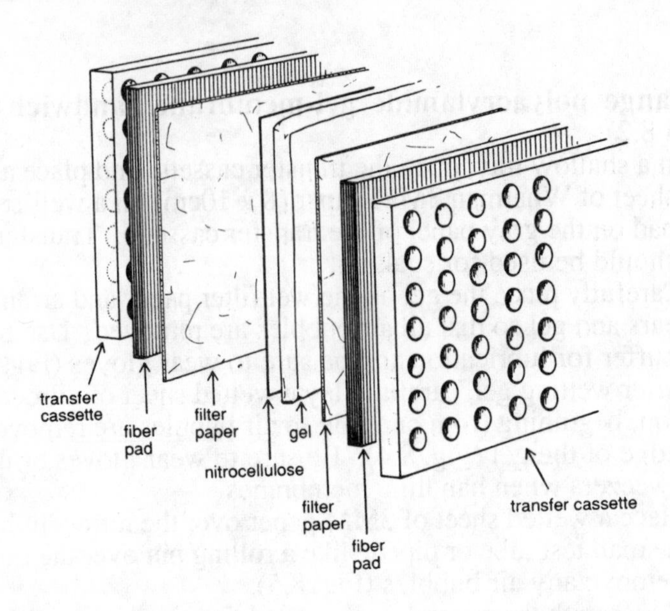

Figure 8.1. Sandwich involving polyacrylamide gel and nitrocellulose membrane for electroblotting.

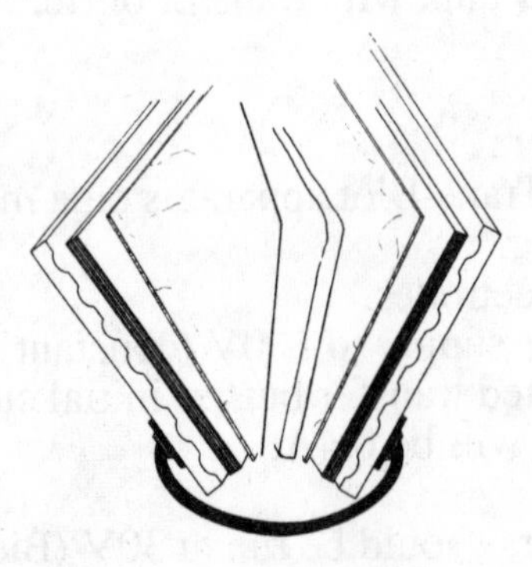

Figure 8.2. Sideview of sandwich in Fig. 8.1.

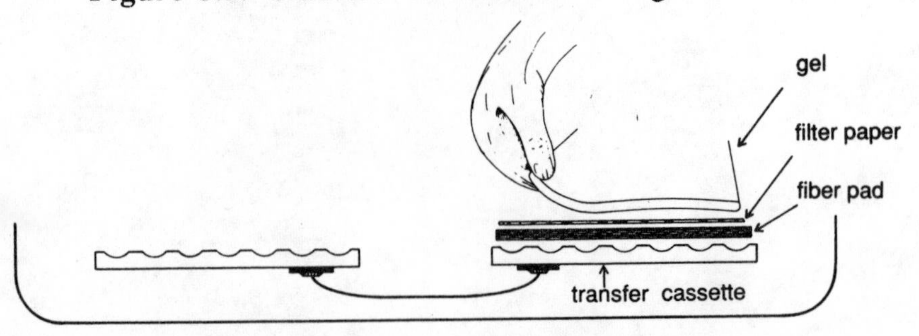

Figure 8.3. Placing the gel on filter paper.

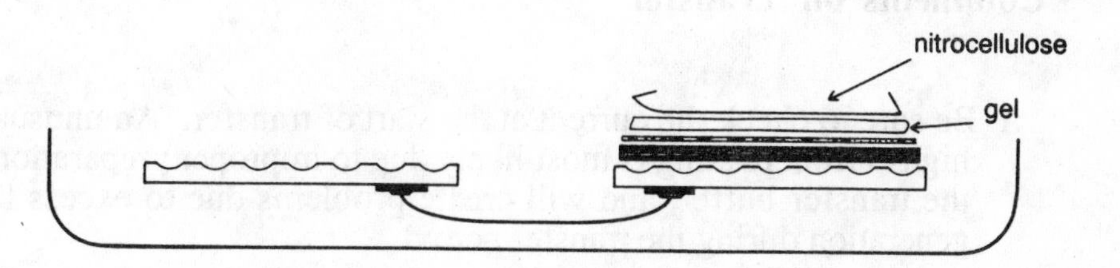

Figure 8.4. Placing nitrocellulose on gel.

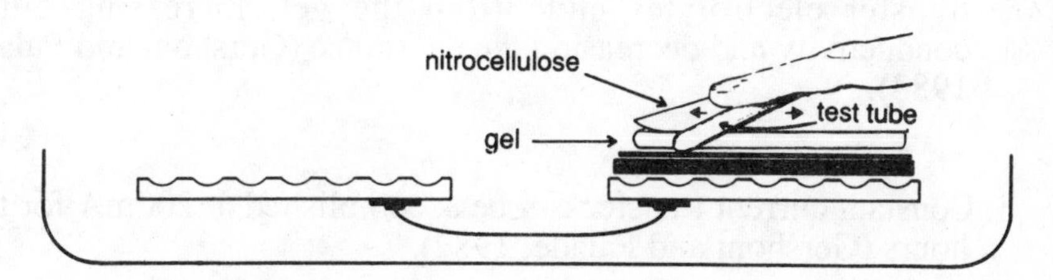

Figure 8.5. Removing air bubbles with a test tube.

Figure 8.6. Inserting transfer cassette into electrode assembly.

• Comments on Transfer

1. Be sure to check the current at the start of transfer. An unusually high current reading is most likely due to improper preparation of the transfer buffer, and will create problems due to excess heat generation during the transfer period.

2. Lower buffer ionic strength allows higher voltage without high current and heat generation (Bers and Garfin, 1985), but during transfer electrolytes elute from the gel, increasing buffer conductivity and decreasing the resistance (Gershoni and Palade, 1983).

3. Constant current transfer can be accomplished at 200mA for two hours (Gershoni and Palade, 1982).

4. It may be helpful to incubate the gel in transfer buffer for 15 - 30 minutes prior to transfer if distortion of bands due to gel swelling during transfer is a problem or if electrolytes eluting from the gel cause an excessive increase in temperature during the transfer (Gershoni and Palade, 1983).

5. The use of pre-stained protein molecular weight markers, and staining of the gel after electroblotting, allow visual determination of the completeness of transfer.

6. For large proteins, a lower acrylamide concentration may result in better transfer (Peluso and Rosenberg, 1987). Another possibility is a two-step elution procedure which allows transfer of low and high molecular weight proteins (Otter et al., 1987).

7. For a guide to transfer conditions for different membranes or transfer buffers, see Bio-Rad Trans-Blot Instruction Manual.

8. The most common problem encountered with electrophoretic transfer of proteins to a membrane is poor protein transfer. This is easily monitored by staining the gel or membrane for proteins after electrotransfer. Four strategies can be tried if the protein of interest binds poorly to nitrocellulose:
   a. Add methanol to 20% in the transfer buffer. Methanol:
      - Decreases efficiency of protein elution from the gel but improves absorption to nitrocellulose (Gershoni and Palade, 1982).
      - Lengthens elution time for large proteins (Gershoni and Palade, 1983) and, due to removal of SDS, can reduce transferability of proteins with a high isoelectric point (Peluso and Rosenberg, 1987 and Bers and Garfin, 1985).
      - Prevents gel swelling (Bers and Garfin, 1985).
   b. Add SDS to 0.1% in the transfer buffer. Low SDS concentrations (around 0.1%) in the transfer buffer can improve transfer efficiency (Towbin and Gordon, 1984), but may reduce the membrane's ability to retain some proteins (Peluso and Rosenberg, 1987).
   c. Use lower pore size nitrocellulose (0.2μm).
   d. Glutaraldehyde cross-link the filter immediately after transfer. Glutaraldehyde cross-linking the proteins immediately after transfer to the membrane improves the stability of small acidic proteins for immunoblotting which may otherwise diffuse out from the membrane. Use 0.2% glutaraldehyde in TBS for 45 minutes (Van Eldik and Wolchok, 1984).

- **Immunodetection**

Once the proteins have been transferred from the polyacrylamide gel to the nitrocellulose membrane, detection of specific proteins proceeds by the use of antibodies. Prior to the addition of antibodies, the membrane is coated with a blocking agent, typically a 3% solution of bovine serum albumin (BSA) in Tris-buffered saline (TBS). Blocking the membrane is important so that antibodies do not bind nonspecifically to the membrane. The first antibody (also called the primary antibody) recognizes the protein of interest while the second antibody recognizes the F_c portion of the first antibody. The second antibody is coupled to an enzyme or another detectable reagent which produces a colored product. The protocol below describes the use of a horseradish peroxidase-conjugated second antibody. Subsequently, modifications of the protocol are provided which describe the use of alkaline phosphatase-, gold-, or ^{125}I-conjugated second antibodies.

Sensitivity:
Horseradish peroxidase: 10 - 20pg (Bers and Garfin, 1985)
Alkaline phosphatase: 10 - 50pg (Bio-Rad Bulletin 1310, 1987)
Immunogold: 1 - 25pg (Bio-Rad Bulletin 1310, 1987)
^{125}I: 50 - 100pg, 1pg with high specific activity ^{125}I (Bers and Garfin, 1985)

Time required: 4 - 5 hours, or incubations may continue overnight.

Solutions to prepare:
- TBS (Tris-buffered saline)
- 3% BSA (Bovine serum albumin) in TBS or 1% nonfat dry milk in TBS
- 0.5% BSA/TBS

TBS (Tris-Buffered Saline), 1 liter

5ml 2M Tris-HCl (pH 7.5)	10mM
37.5ml 4M NaCl	150mM
957.5ml distilled water	

Can also be made as a 10x solution.

All steps can be carried out at room temperature or at 4°C.

1. **Block membrane:**
   a. Disconnect transfer apparatus, remove transfer cassette, and peel 3MM paper from nitrocellulose.
   b. Using forceps or wearing gloves, remove nitrocellulose membrane from transfer apparatus to a small container or to a Seal-A-Meal bag.
   c. Add at least 8ml 3% BSA/TBS (enough to cover the membrane).
   d. Rock the filter gently for 30 minutes to 1 hour, making sure that the entire filter is in contact with the BSA solution. The blocked membrane may be stored in the BSA/TBS solution with 1mM NaN_3 overnight or for as long as several days depending on the stability of the protein.

2. **Wash membrane:** Pour off BSA solution and rinse briefly with TBS three times.

3. **First antibody wash:**
   a. Pour off TBS, add first antibody at appropriate dilution in 8ml 0.5% BSA/TBS (see comment 2.d. below for dilutions).
      To conserve volume, place the membrane in a Seal-A-Meal bag, add first antibody solution and seal. It may be possible to use a 4ml of solution for a 6cm x 8cm membrane. Be careful to remove air bubbles as completely as possible. Air bubbles prevent antibody-epitope contact and can result in bands that remain undetected.
   b. Rock gently for at least 1 hour. Overnight incubations are possible and may increase detection sensitivity.

4. **Wash membrane:**
   a. Pour off first antibody solution from membrane.
   b. Wash twice for 10 minutes with TBS.

5. **Second antibody wash:**
   a. Pour off TBS.
   b. Add second antibody at appropriate dilution in 8ml 0.5% BSA/TBS.
   c. Rock the membrane gently for at least an hour. An overnight incubation is acceptable.

6. **Wash membrane:**
   a. Pour off second antibody solution from membrane.
   b. Rinse for 30 minutes with TBS, with 3 changes. At this time, if you are incubating the membrane in a Seal-A-Meal bag, it is advisable to remove membrane to a glass or plastic container.

7. **Develop nitrocellulose** (horseradish peroxidase):
   a. Prepare developing reagent:
      1ml chloronaphthol solution (30mg/ml in methanol)
      add 10ml methanol
      add room temperature TBS to 50ml
      add 30μl 30% hydrogen peroxide (H_2O_2)
   b. Pour off TBS from membrane, add developing reagent.
   c. Rock nitrocellulose gently, monitoring development.
   d. If a heavy precipitate forms, replace developing reagent with a freshly made solution.
   e. Development should be complete in 5 to 30 minutes.
   f. Stop development by washing membrane with distilled water for 30 minutes with 3 changes.
   g. Dry nitrocellulose with absorbant paper and store the membrane protected from light and the atmosphere. A plastic envelope in a notebook is usually suitable.
   h. Photograph within a week, since the signal may fade with time and the nitrocellulose may begin to turn yellow. A typical pattern is presented in Fig. 8.7.

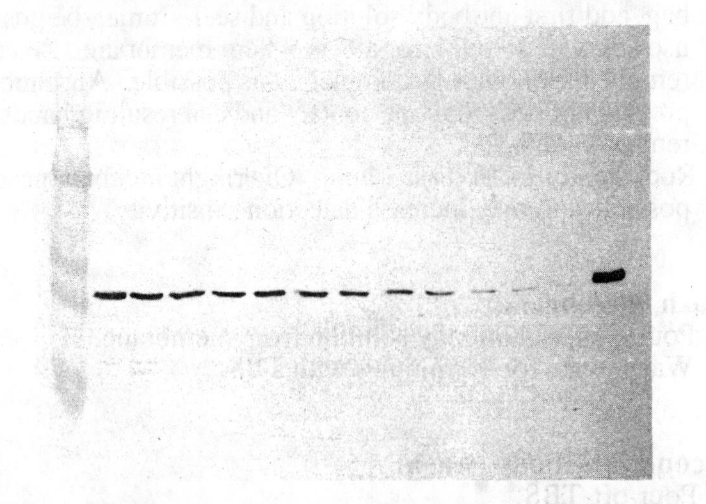

Figure 8.7. Immunoblot detecting yeast tubulin with a tubulin-specific first antibody and subsequent staining with alkaline phosphatase. Pairs of samples were run in a 2-fold dilution series from 20μg to 0.62μg. The far left lane contains molecular weight markers and the far right lane contains a brain tubulin marker.

• Comments on Immunodetection

1. Membrane Blocking and Washing

a. Nitrocellulose may be baked for 1 hour at 80°C in a vacuum oven following transfer to help stabilize protein binding (Hsu, 1984).

b. Step 1: Blocking can also be accomplished with 3% gelatin in TBS. Blocking of nylon membranes should be done with 10% BSA at 45 - 50°C for at least 12 hours.

c. Step 2: After this step, the filter may be dried and stored. Rewet filter in 0.5% BSA/TBS before continuing with immunodetection protocol.

d. Bovine serum albumin (BSA) is most commonly used to reduce background in the blocking and washing steps. BSA is often contaminated with immunoglobulin G, which makes it unsatisfactory for use with protein A (Bers and Garfin, 1985). Use Tween-20 or gelatin instead.

e. Other blocking agents include:
   - BLOTTO: 5% (w/v) nonfat dry milk, 0.01% Antifoam A, Sigma (Johnson et al., 1984).
   - 3% gelatin for blocking, 1% for antibody incubations (Bio-Rad Immun-Blot (GAR-HRP) Assay Kit Bulletin).
   - 0.05% Tween-20 for blocking and antibody incubations (Bers and Garfin, 1985).

f. BLOTTO may give inconsistent results with different batches of dry milk, and it is possible that blocking with milk will give a higher background than with BSA (Kaufmann et al., 1987). In addition, these products may contain immunoglobulin G which will become a problem if protein A is used for detection, and they may react with lectin probes such as concanavalin A (Bers and Garfin, 1985). BLOTTO may be omitted from the washing steps if it has already been used for blocking the membrane (Peluso and Rosenberg, 1987).

g. Congealing of gelatin at low temperatures may be a problem (Bers and Garfin, 1985).

2. Use of Antibodies

a. Steps 3 and 5: For routine use, prepare convenient aliquots of the first and second antibodies and store them in the freezer.

b. Color development is strongest at high first and second antibody concentrations. However, if antibody concentrations are too high, nonspecific background bands will appear unless the enzyme reaction is stopped promptly. See Bers and Garfin (1985) for additional troubleshooting advice.

c. Antibodies can be diluted in 0.5% BSA/TBS and stored at -20°C (Burnette, 1981). Repeated freezing and thawing can lead to antibody aggregation and loss of activity.

d. Antibody dilution guidelines:
- First antibody: extremely variable, from 1:10 to 1:100,000.
- Second antibody: commonly 1:500 to 1:4000.

3. General Immunodetection Comments

a. Protein transfer is not the same on both sides of the nitrocellulose membrane (as becomes evident if using pre-stained molecular weight markers). Be sure to look at the signal on both sides of the immunoblot during development.

b. SDS-PAGE may cause a loss of antigenicity, especially for monoclonal antibody detection (Burnette, 1981).

c. Possible reasons for poor color development:
- First or second antibody is inactive or non-saturating, due to improper storage or excess dilution. A control for first antibody activity may involve spotting the antigen directly on a piece of nitrocellulose and performing the immunoassay. Similarly, the second antibody can be tested for binding to a different first antibody produced in the same species.
- Insufficient antigen on the nitrocellulose membrane. Stain the nitrocellulose for total protein or include a known amount of control antigen on the blot.

d. Possible reasons for high background:
- Insufficient washing between antibody incubations.
- Insufficient blocking.
- Contaminated fiber pads or transfer buffer.
- Antibody concentrations too high. Include 0.05% Tween 20 in antibody buffers.

Additional suggestions for solving problems due to diffuse or specific background bands may be found in Harlow and Lane (1988, p. 510).

4. Horseradish Peroxidase

a. Horseradish peroxidase-coupled second antibody yields a blue-purple color when reacted with chloronaphthol as described above. Alternative development substrates include aminoethylcarbazole (yielding a red color) and diaminobenzidine (yielding a brown color) which may provide greater sensitivity (see Harlow and Lane, 1988).

b. After transferring the second antibody aliquot to the BSA/TBS solution in step 5, keep the tube containing a trace of the second antibody for testing in step 7. Prepare the developing reagent and test the reagent by adding 50 - 100µl to the used second antibody tube. If the solution does not start to turn blue within 2 minutes, there is a problem with the developing reagent.

c. Chloronaphthol stock solution (30mg/ml in methanol) can be made and stored at -20°C for at least a year.

d. Although the horseradish peroxidase detection system is economical and rapid to perform, some possible disadvantages include:
- Fading with exposure to light
- Non-specific reaction by endogenous peroxidase enzymes on immunoblot

e. Possible reasons for poor color development:
- Developing reagent is inactive, especially if H_2O_2 is inactive or chloronaphthol has precipitated out of solution. Prepare new developing reagent, using fresh enzyme immunoassay (EIA) grade chemicals.
- Horseradish peroxidase may be inactivated by azide or impure methanol.

D. Protocol Modifications for Other Detection Methods

• Alkaline Phosphatase

Alkaline phosphatase-conjugated second antibody is reacted with a bromochloroindolyl phosphate - nitro blue tetrazolium substrate to give a dark purple precipitate. An important advantage over the horseradish peroxidase reaction is that the colored alkaline phosphatase product is stable and will not fade.

Follow Immunodetection Protocol (Section C) through step 6b, substituting an alkaline phosphatase-conjugated second antibody in step 5b.

7. Development (from Harlow and Lane, 1988)

a. Prepare developing reagents

Reagent Solutions

(can be prepared in advance and stored at 4°C for over a year)

0.1M Tris-HCl (pH 9.5), 0.1M NaCl, 5mM $MgCl_2$
(Alkaline Phosphatase Buffer)

50mg/ml 5-bromo-4-chloro-3-indolyl phosphate in 100% dimethylformamide (BCIP Solution)

50mg/ml *p*-nitro blue tetrazolium chloride in 70% dimethylformamide (NBT Solution)

b. Wash nitrocellulose for 5 minutes in Alkaline Phosphate Buffer.

c. Prepare Developing Reagent:

66µl NBT Solution
10ml Alkaline Phosphatase Buffer
mix well
add 33µl BCIP Solution
Use within 1 hour

d. Pour off Alkaline Phosphatase Buffer and add 10ml Developing Reagent to nitrocellulose.

e. Incubate at room temperature or at 37°C to speed reaction.

f. Reaction is mostly complete within 30 minutes but can be permitted to continue overnight to increase signal, and can be stopped by rinsing filter with 20mM EDTA in TBS.

• Immunogold

Colloidal gold conjugates give a stable signal after development and are very easy and rapid to use. Unfortunately, high sensitivity requires a more complicated enhancement step following initial development.

Follow Immunodetection Protocol through step 4b.

5. Development (from Hsu, 1984)

   a. Prepare gold-conjugated second antibody by diluting to A_{525}=0.5 in 0.05% Tween 20 in Tris-Buffered Saline (TBS).
   b. Remove TBS from nitrocellulose membrane and add second antibody solution.
   c. Incubate for 30 - 60 minutes.
   d. Rinse membrane with water after development.

6. Silver Enhancement of Immunogold Signal (Brada and Roth, 1984)

   a. After development (step 5c), rinse membrane twice for 5 minutes in TBS.
   b. Rinse membrane for 1 minute in distilled water.
   c. Dip nitrocellulose membrane into developer solution for 1 - 2 minutes (see Danscher, 1981).
   d. Wash membrane in tap water for 1 minute.
   e. Incubate membrane for 10 minutes in photographic fixer.

Color produced with gold stain is red, subsequent silver enhancement produces a black color (Brada and Roth, 1984).

A brief centrifugation just before use (250 - 4000xg) removes aggregated gold particles. 5nm gold particles give more sharply focused bands than 15nm particles (Surek and Latzko, 1984).

• ^{125}I Methods

^{125}I detection systems are in disfavor due to handling and disposal problems, short half-lives, long wash steps to remove background, long development times and high cost (Bio-Rad Bulletin 1310, 1987). Protein A is less desirable than a 2nd antibody, since it does not recognize IgGs of all species or IgG subtypes. In addition, its binding is not polyvalent (Allen et al., 1984).

Follow Immunodetection Protocol through step 6b, substituting ^{125}I-labeled second antibody or ^{125}I-Protein A. **All solutions used after addition of the radiolabeled protein must be handled as radioactive waste.** 5 - 10μCi (Kaufmann et al., 1987) or 2-5x10^{6}cpm (Renart and Sandoval, 1987) of ^{125}I-labeled probe have been used.

7. Detection (from Kaufmann et al., 1987)

   a. Dry nitrocellulose thoroughly. Radioactive ink may be used to mark corners or molecular weight markers on the membrane.
   b. Wrap the membrane in plastic wrap.
   b. Expose to Kodak X-Omat XAR-5 or XRP-5 film with an intensifying screen at -70°C.
   c. Develop film according to instructions.

E. Staining for Total Protein

Staining for total protein on a membrane may be useful for monitoring the efficiency of transfer or for identifying an immunochemically detected band. For example, it is better to monitor the transfer efficiency in one or two lanes before probing the rest of the lanes with expensive antibody reagents. Three widely used methods for staining of nitrocellulose membranes are amido black, India ink, and Ponceau S staining. Nonspecific binding of anionic dyes makes these three methods less satisfactory for nylon membranes, but a biotin-avidin-horseradish peroxidase stain should provide satisfactory results (Bio-Rad Bulletin 1310, 1987). Colloidal gold staining provides the highest sensitivity.

Detection Levels:
- Amido black - 30ng per band (Bers and Garfin, 1985)
- India ink - 6ng per band (Glenney, 1986)
- Enhanced Colloidal Gold - 400pg per band (Bio-Rad Bulletin 1310, 1987)
- Biotin - 10 - 50ng per band (Bio-Rad Bulletin 1310, 1987)

• **Amido Black Staining** of nitrocellulose membranes:
1. Stain 1 minute in 0.1% amido black 10B/25% isopropanol/10% acetic acid.
2. Destain for 30 minutes in 25% isopropanol/10% acetic acid.
3. Wash filters in TBS or water before drying to reduce shrinkage (Gershoni and Palade, 1982).

• **India Ink Staining** of nitrocellulose membranes:
1. Wash membrane twice for 5 minutes in 0.5% Tween 20/TBS.
2. Stain for at least 2 hours in 1μl india ink/ml distilled water.
3. Destain by rinsing several times for 5 minutes with distilled water. (Glenney, 1986).

• **Ponceau S Staining** of nitrocellulose membranes:
1. Stain for 5 - 10 minutes in 0.2% Ponceau S/3% TCA/3% sulfosalicylic acid
2. Wash away stain with TBS.
3. Ponceau S staining can be followed by immunodecoration (Harlow and Lane, 1988).

• **Comments on Total Protein Staining**

1. Staining with anionic dyes in methanolic solutions will cause some shrinking of the nitrocellulose, so India ink staining may be preferred if exact replicas of blots are required.

2. No detectable quenching of radiolabel was observed due to India ink staining (Glenney, 1986).

3. Certain protein bands are differentially stained with different staining methods. For example, acidic proteins stain only at pH 3.5 with colloidal gold (Rohringer and Holden, 1985).

4. Ponceau S staining is not very sensitive and the red color is difficult to photograph; however, the stain may be washed away and immunodetection may follow the staining.

F. Erasing Immunoblots

Erasing a blot involves removal of the primary and secondary antibodies. This permits repetitive use of a single blot. It should be noted that some epitopes will be damaged by the erasure treatment.

Erasing Buffer, 100ml

6.25ml 1M Tris-Cl (pH 6.8)	62.5mM
20ml 10% SDS	2%
0.7ml 2-Mercaptoethanol	100mM
73ml H_2O	

- **Steps**

1. After development of 2nd antibody reaction or prior to drying for autoradiography, incubate nylon or nitrocellulose membrane in 5% powdered nonfat milk/TBS for 10 minutes at room temperature.

2. Dry on absorbent paper at room temperature. After 5 minutes, move to a fresh piece of absorbent paper to prevent membrane from sticking.

3. To erase, incubate the dried membrane for 30 minutes at 70°C in Erasing Buffer.

4. Following two 10 minute washes in TBS, nitrocellulose membranes should be incubated with 5% nonfat dry milk/TBS for 6 hours and nylon membranes should be recoated with 10% nonfat dry milk for 6 - 8 hours prior to reprobing with a new first antibody (Kaufmann et al., 1987).

- **Alternative erasing conditions**: see Renart and Sandoval (1984), Earnshaw and Rothfield (1985), and Surek and Latzko (1984).

III. Discussion

A. Membrane Storage

- To reduce tendency of nitrocellulose membranes to stick to absorbent paper when drying: first incubate the blots for 10 minutes in nonfat dry milk/TBS, then dry 5 minutes and move to fresh absorbent paper (Kaufmann et al., 1987).

- Nitrocellulose paper can be stored for as long as a year following transfer before probing (Gershoni and Palade, 1982).

B. Tranfer Anomalies

- Sometimes, proteins do not electroelute efficiently because they are fortuitously at their isoelectric point. In this case, other buffer conditions should be used (Gershoni and Palade, 1983). Strongly basic proteins such as histones, lysozymes or cytochromes may transfer poorly (Szewczyk and Kozloff, 1985).

- For isoelectric focusing gels, SDS-urea gels or nondenaturing gels containing basic proteins, transfer in 0.7% acetic acid with **gel cassette in reverse orientation** (that is, transferring toward the cathode).

- Exceeding the membrane binding capacity may reduce the signal in subsequent detection steps (Gershoni and Palade, 1983).

C. References for Other Uses of Immunoblots

- **Epitope Mapping:** Glenney et al., 1983; Mendelson et al., 1984.

- **Structural Domain Analysis:** Russel et al., 1984; Yurchenco et al., 1982.

- **Dot Blot:** May be applied to analysis of column fractions, sucrose gradients or pulse-chase experiments (Hawkes et al., 1982; Bosman et al., 1983; Talbot et al., 1984).

- **Functional Assay:**

Although most functional tests for electroblotted proteins currently involve ligand binding assays (see section below), improved protein renaturation techniques should allow the establishment of other specific enzymatic assays. Enzymatic assays have been described for proteins in polyacrylamide gels, including dehydrogenases, phosphatases, esterases, oxidases, and peptidases, and these are referenced in Chapter 6, section III.B.

Conditions for preserving or restoring enzyme activity during electroblotting may involve sample treatment prior to electrophoresis, gel incubation prior to electrotransfer, or incubation of the membrane after protein transfer. Protein sample treatments before and during gel electrophoresis include eliminating the use of sulfhydryl reagents (Islan et al., 1983; Daniel et al., 1983), EDTA (Gershoni et al., 1983), or sample heating (Daniel et al., 1983). Polyacrylamide gel treatment may involve washing the gel in a buffered solution to remove SDS (Bowen et al., 1980; Wolff et al., 1985). Nitrocellulose membranes have been incubated in solutions containing low amounts of detergent to allow protein renaturation (Haeuptle et al., 1983).

- **Ligand Binding:**

A wide variety of ligands have been used as probes for proteins after immunoblotting, including:

DNA (Bowen et al., 1980; Hoch, 1982; Patel and Cook, 1983)

RNA (Bowen et al., 1980; Rozier and Mache, 1984)

Lectins (Hawkes, 1982)

Hormones (Haeuptle et al., 1983)

Toxins (Gershoni and Palade, 1983)

Viruses (Co et al, 1985)

Heparin (Cardin et al., 1984)

Calmodulin (Gershoni and Palade, 1983; Flanagan and Yost, 1984)

Histones (Bowen et al., 1980)

Whole Cells (Hayman et al., 1982)

GTP (McGrath et al., 1984)

Calcium (Fong et al., 1988)

Zinc (Serrano et al., 1988)

- **Purifying Antibody From an Immunoblot:** Smith and Fisher, 1984; Olmsted, 1981.

- **Cutting Protein Bands From a Nitrocellulose Membrane for Antibody Production:** Knudson, 1985; Harlow and Lane, 1988, p. 498.

- **Protein Identification: Amino Acid Analysis and Protein Sequencing**

Minute amounts of protein (as little as 10pmol) may be analyzed for amino acid composition (Tous et al., 1989) or sequence (Matsudaira, 1987) after electroblotting on polyvinylidene difluoride (PVDF) membranes. Transferred protein or peptide bands are stained with Coomassie blue and then excised from the membrane.

IV. Suppliers

Electroblotting Apparatus: Bio-Rad Mini Trans-Blot Electrophoretic Transfer Cell; Hoefer Model TE 22 Mighty Small Transfer Unit; Schleicher and Schuell Mini Transfer System

Power supply with a capacity of 200V, 0.6A: Bio-Rad, Pharmacia

Filter Paper: Bio-Rad; Schleicher and Schuell; Whatman

Nitrocellulose: Bio-Rad; Millipore; Schleicher and Schuell

Nylon: Bio-Rad (Zeta-probe); CUNO (Zetabind)

Rocker: Hoefer Red Rocker

India Ink: Pelikan fount india drawing ink, Pelikan AG; Speedball drawing ink, dense black india, No. 3211, Hunt Mfg. Co.

Ponceau S (3-hydroxy-4-[2-sulfo-4-(sulfo-phenylazo)phenylazo]-2,7-naphthalene disulfonic acid: Sigma

Colloidal gold and biotin total protein staining kits: Bio-Rad

Horseradish Peroxidase Conjugates: Bio-Rad; Calbiochem; Sigma

Alkaline Phosphatase Conjugates: Bio-Rad; Calbiochem; Sigma

Colloidal Gold Conjugates: Bio-Rad; Sigma

V. References

Allen, R.C., C.A. Saravis, and H.R. Maurer. 1984. Gel Electrophoresis and Isoelectric Focusing of Proteins. Selected Techniques. Walter de Gruyter, Berlin. pp. 221-230.

Bers, G. and D. Garfin. 1985. BioTechniques 3: 276-288. Protein and Nucleic Acid Blotting and Immunobiochemical Detection.

Bio-Rad Bulletin 1310. 1987. Western Blotting Detection Systems: How do you choose?

Bio-Rad Immun-Blot (GAR-HRP) Assay Kit Bulletin.

Bio-Rad Mini Trans-Blot Electrophoretic Transfer Cell Instruction Manual.

Blake, M.S., K.H. Johnston, G.J. Russel-Jones, and E.C. Gotschlich. 1984. Anal. Biochem. 136: 175-179. A Rapid, Sensitive Method for Detection of Alkaline Phosphatase-Conjugated Anti-Antibody on Western Blots.

Bosman, F.T., G. Cramer-Knijnenburg, and J.v.B. Henegouw. 1983. Histochemistry 77: 185-194. Efficiency and Sensitivity of Indirect Immunoperoxidase Methods.

Bowen, B., J. Steinberg, U.K. Laemmli, and H. Weintraub. 1980. Nuc. Acids Res. 8: 1-20. The Detection of DNA-Binding Proteins by Protein Blotting.

Brada, D. and J. Roth. 1984. Anal. Biochem. 142: 79-83. "Golden Blot" - Detection of Polyclonal and Monoclonal Antibodies Bound to Antigens on Nitrocellulose by Protein A - Gold Complexes.

Burnette, W.N.. 1981. Anal. Biochem. 112: 195-203. "Western Blotting": Electrophoretic Transfer of Proteins From Sodium Dodecyl Sulfate - Polyacrylamide Gels to Unmodified Nitrocellulose and Radiographic Detection With Antibody and Radioiodinated Protein A.

Cardin, A.D., K.R. Witt, and R.L. Jackson. 1984. Anal. Biochem. 137: 368-373. Visualization of Heparin-Binding Proteins by Ligand Blotting with ^{125}I-Heparin.

Co, M.S., G.N. Gaulton, B.N. Fields, and M.I. Greene. 1985. Proc. Natl. Acad. Sci. USA 82: 1494-1498. Isolation and Biochemical Characterization of the Mammalian Reovirus Type 3 Cell-Surface Receptor.

Daniel, T.O., W.J. Schneider, J.L. Goldstein, and M.S. Brown. 1983. J. Biol. Chem. 258: 4606-4611. Visualization of Lipoprotein Receptors by Ligand Blotting.

Danscher, G. 1981. Histochemistry 71: 81-88. Localization of Gold in Biological Tissue.

Earnshaw, W.C. and N. Rothfield. 1985. Chromosoma 91: 313-321. Identification of a Family of Human Centromere Proteins Using Autoimmune Sera from Patients with Scleroderma.

Flanagan, S.D. and B. Yost. 1984. Anal. Biochem. 140: 510-519. Calmodulin-Binding Proteins: Visualization by ^{125}I-Calmodulin Overlay on Blots Quenched with Tween 20 or Bovine Serum Albumin and Poly(ethylene oxide).

Fong, K.C., J.A. Babitch, and F.A. Anthony. 1988. Biochim. Biophys. Acta 952: 13-19. Calcium Binding to Tubulin.

Gershoni, J.M., E. Hawrot, and T.L. Lentz. 1983. Proc. Natl. Acad. Sci. USA 80: 4973-4977. Binding of Alpha-Bungarotoxin to Isolated Alpha Subunit of the Acetylcholine Receptor of *Torpedo californica*: Quantitative Analysis With Protein Blots.

Gershoni, J.M. and G.E. Palade. 1982. Anal. Biochem. 124: 396-405. Electrophoretic Transfer of Proteins From Sodium Dodecyl Sulfate - Polyacrylamide Gels to a Positively Charged Membrane Filter.

Gershoni, J.M. and G.E. Palade. 1983. Anal. Biochem. 131: 1-15. Protein Blotting: Principles and Application.

Glenney, J. 1986. Anal. Biochem. 156: 315-319. Antibody Probing of Western Blots Which Have Been Stained with India Ink.

Glenney, Jr., J.R., P. Glenney and K. Weber. 1983. J. Mol. Biol. 167: 275-293. Mapping the Fodrin Molecule with Monoclonal Antibodies.

Haeuptle, M.-T., M.L. Aubert, J. Djiane, and J.-P. Kraehenbuhl. 1983. J. Biol. Chem. 258: 305-314. Binding Sites for Lactogenic and Somatogenic Hormones from Rabbit Mammary Gland and Liver.

Harlow, E. and D. Lane. 1988. Antibodies: A Laboratory Manual. 726 pages. Cold Spring Harbor Laboratory, Cold Spring Harbor, New York.

Hawkes, R. 1982. Anal. Biochem. 123: 143-146. Identification of Concanavalin A-Binding Proteins after Sodium Dodecyl Sulfate - Gel Electrophoresis and Protein Blotting.

Hawkes, R., E. Niday, and J. Gordon. 1982. Anal. Biochem. 119: 142-147. A Dot-Immunobinding Assay for Monoclonal and Other Antibodies.

Hayman, E.G., E. Engvall, E. A'Hearn, D. Barnes, M. Pierschbacher, and E. Ruoslahti. 1982. J. Cell Biol. 95: 20-23. Cell Attachment on Replicas of SDS Polyacrylamide Gels Reveals Two Adhesive Plasma Proteins.

Hoch, S.D. 1982. Biochem. Biophys. Res. Comm. 106: 1353-1358. DNA-Binding Domains of Fibronectin Probed Using Western Blots.

Hsu, Y.-H. 1984. Anal. Biochem. 142: 221-225. Immunogold for Detection of Antigen on Nitrocellulose Paper.

Islan, M.N., R. Briones-Urbina, G. Bako, and N.R. Farid. 1983. Endocrin. 113: 436-438. Both TSH and Thyroid-stimulating Antibody of Graves' Disease Bind to an M_r 197,000 Holoreceptor.

Johnson, D.A., J.W. Gautsch, J.R. Sportsman, and J.H. Elder. 1984. Gene Analysis Techniques 1: 3-8. Improved Technique Utilizing Nonfat Dry Milk for Analysis of Proteins and Nucleic Acids Transferred to Nitrocellulose.

Kaufmann, S.H., C.M. Ewing, and J.H. Shaper. 1987. Anal. Biochem. 161: 89-95. The Erasable Western Blot.

Knudson, K.A. 1985. Anal. Biochem. 147: 285-288. Proteins Transferred to Nitrocellulose for Use as Antigens.

McGrath, J.P., D.J. Capon, D.V. Goeddel, and A.D. Levinson. 1984. Nature 310: 644-649. Comparative Biochemical Properties of Normal and Activated Human *ras* p21 Protein.

Matsudaira, P. 1987. J. Biol. Chem. 262: 10035-10038. Sequence from Picomole Quantities of Proteins Electroblotted onto Polyvinylidene Difluoride Membranes.

Mendelson, E., B.J. Smith, and M. Bustin. 1984. Biochem. 23: 3466-3471. Mapping the Binding of Monoclonal Antibodies to Histone H5.

Olmsted, J.B. 1981. J. Biol. Chem. 256: 11955-11957. Affinity Purification of Antibodies from Diazotized Paper Blots of Heterogeneous Protein Samples.

Otter, T., S.M. King, and G.B. Witman. 1987. Anal. Biochem. 162: 370-377. A Two-Step Procedure for Efficient Electrotransfer of Both High-Molecular-Weight (>400,000) and Low-Molecular-Weight (<20,000) Proteins.

Patel, S.B. and P.R. Cook. 1983. EMBO J. 2: 137-142. The DNA-Protein Cross: A Method for Detecting Specific DNA-Protein Complexes in Crude Mixtures.

Peluso, R.W. and G.H. Rosenberg. 1987. Anal. Biochem. 162: 389-398. Quantitative Electrotransfer of Proteins fron Sodium Dodecyl Sulfate-Polyacrylamide Gels onto Positively Charged Nylon Membranes.

Renart, J. and I.V. Sandoval. 1984. Meth. Enzymol. 104: 455-459. Western Blots.

Rohringer, R. and D.W. Holden. 1985. Anal. Biochem. 144: 118-127. Protein Blotting: Detection of Proteins with Colloidal Gold, and of Glycoproteins and Lectins with Biotin-Conjugated and Enzyme Probes.

Rozier, C. and R. Mache. 1984. Nuc. Acids Res. 12: 7293-7304. Binding of 16S rRNA to Chloroplast 30S Ribosomal Proteins Blotted on Nitrocellulose.

Russel, D.W., W.J. Schneider, T. Yamamoto, K.J. Luskey, M.S. Brown, and J.L. Goldstein. 1984. Cell 37: 577-585. Domain Map of the LDL Receptor: Sequence Homology with the Epidermal Growth Factor Precursor.

Serrano, L., J.E. Dominguez, and J. Avila. 1988. Anal. Biochem. 172: 210-218. Identification of Zinc-Binding Sites of Proteins: Zinc Binds to the Amino-Terminal Region of Tubulin.

Smith, D.E. and P.A. Fisher. 1984. J. Cell Biol. 99: 20-28. Identification, Developmental Regulation, and Response to Heat Shock of Two Antigenically Related Forms of a Major Nuclear Envelope Protein in Drosophila Embryos: Application of an Improved Method for Affinity Purification of Antibodies Using Polypeptides Immobilized on Nitrocellulose Blots..

Surek, B. and E. Latzko. 1984. Biol. Biochem. Res. Comm. 121: 284-289. Visualization of Antigenic Proteins Blotted onto Nitrocellulose Using the Immuno-Gold-Staining (IGS)-Method.

Szewczyk, B. and L.M. Kozloff. 1985. Anal. Biochem. 150: 403-407. A Method for the Efficient Blotting of Strongly Basic Proteins from Sodium Dodecyl Sulfate-Polyacrylamide Gels to Nitrocellulose

Talbot, P.J., R.L. Knobler, and M.J. Buchmeier. 1984. J. Immunol. Methods 73: 177-188. Western and Dot Immunoblotting Analysis of Viral Antigens and Antibodies: Application to Murine Hepatitis Virus.

Tous, G.I., J.L. Fausnaugh, O. Akinyosoye, H. Lackland, P. Winter-Cash, F.J. Vitorica, and S. Stein. 1989. Anal. Biochem. 179: 50-55. Amino Acid Analysis on Polyvinylidene Difluoride Membranes.

Towbin, H. and J. Gordon. 1984. J. Immunol. Meth. 72: 313-340. Immunoblotting and Dot Immunoblotting - Current Status and Outlook.

Towbin, H., T. Staehelin, and J. Gordon. 1979. Proc. Nat. Acad. Sci. 76: 4350-4354. Electrophoretic Transfer of Proteins from Polyacrylamide Gels to Nitrocellulose Sheets: Procedure and Some Applications.

Van Eldik, L.J. and S.R. Wolchok. 1984. Biochem. Biophys. Res. Comm. 124: 752-759. Conditions for Reproducible Detection of Calmodulin and S100B in Immunoblots.

Wolff, P., R. Gilz, J. Schumacher, and D. Riesner. 1985. Nucl. Acids Res. 13: 355-367. Complexes of Viroids with Histones and Other Proteins.

Yurchenco, P.D., D.W. Speicher, J.S. Morrow, W.F. Knowles, and V.T. Marchesi. 1982. J. Biol. Chem. 257: 9102-9107. Monoclonal Antibodies as Probes of Domain Structure of the Spectrin a Subunit.

Appendix 1

Molecular Weights of Commonly Used Chemicals

Chemical	Molecular Weight	Molarity
ACES	182.2	
Acetate (Na salt)	82.0	
Acetic Acid, glacial	60.05	17.4
Acetone	58.1	
Acrylamide	71.1	
Adenosine Triphosphate (ATP, disodium salt)	605.2	
β-Alanine	89.1	
Amido Black 10B	616.5	
Ammonium Hydroxide (NH_4OH)	35.0	14.5 (30%)
Ammonium Persulfate	228.2	
Ammonium Sulfate [$(NH_4)_2SO_4$]	132.1	
Aprotinin	~6500	
Barbital (barbituric acid)	128.1	
Bicinchoninic Acid (BCA)	420.5	
Bicine	163.2	
Bis-Acrylamide (N, N' methylenebis-acrylamide)	154.2	
Boric Acid	61.8	
5-Bromo-4-Chloro-3-Indolyl Phosphate (BCIP)	348.7	
Bromophenol Blue (sodium salt)	692.0	
Cacodylate (sodium salt trihydrate)	214.0	
CAPS	221.3	
CHAPS	614.9	
CHES	207.3	
Chloroform	119.4	
4-Chloro-1-naphthol	178.6	
Citric Acid	192.1	
Coomassie Blue R-250	826.0	
Copper Sulfate ($CuSO_4$)	159.6	
Deoxycholate (DOC)	392.6	
Dimethylformamide (DMF)	73.1	

Chemical	Molecular Weight	Molarity
Dithiothreitol (DTT)	154.3	
Ethanol	46.1	
Ethylene diamine tetraacetic acid (EDTA)	292.2	
Ethylene bis(oxyethylenenitrilo)- tetraacetic acid (EGTA)	380.35	
Ethylene Glycol	62.1	
Formaldehyde	30.0	12.1 (37%)
Formate (sodium salt)	68.0	
Formic Acid	46.0	23.4
Glutaraldehyde	100.1	
Glycerol	92.1	
Glycine	75.1	
Glycine-HCl	111.5	
Glycine-NaOH	97.1	
Glycylglycine	132.1	
Guanidine Hydrochloride	95.5	
HEPES	238.3	
Hydrochloric Acid	36.5	12.1 (36.5-38%)
Hydrogen Peroxide (H_2O_2)	34.0	8.8 (30%)
Isopropanol	60.1	
Isopropyl-β-D-thiogalactopyranoside (IPTG)	238.3	
Leupeptin	493.6	
Lithium Dodecyl Sulfate (LiDS)	272.3	
Magnesium Chloride ($MgCl_2$)	95.2	
2-Mercaptoethanol	78.1	14.4
MES	195.2	
Methanol	32.0	
MOPS	209.3	
Nitric Acid	63.0	16 (70%)
p-Nitro Blue Tetrazolium Chloride (NBT)	817.6	
Pepstatin A	686	
Phenyl Methyl Sulfonyl Fluoride (PMSF)	174.2	
Phosphoric Acid	98.0	14.7 (85%)
PIPES	302.4	
Potassium Chloride (KCl)	74.6	
Potassium Hydroxide (KOH)	56.1	
Riboflavin	376.4	
Silver Nitrate ($AgNO_3$)	169.9	
Sodium Azide (NaN_3)	65.0	
Sodium Bicarbonate ($NaHCO_3$)	84.0	

Chemical	Molecular Weight	Molarity
Sodium Carbonate (Na_2CO_3)	106.0	
Sodium Chloride (NaCl)	58.4	
Sodium Citrate ($Na_3C_6H_5O_7$ [$\cdot 2H_2O$])	294.1	
Sodium Dodecyl Sulfate (SDS)	288.4	
Sodium Hydroxide (NaOH)	40.0	
Sodium Phosphate, dibasic (Na_2HPO_4)	142.0	
Sodium Phosphate, monobasic (NaH_2PO_4)	120.0	
Sodium Tartrate ($Na_2C_4H_4O_6$ [$\cdot 2H_2O$])	230.1	
Succinate (free acid)	118.1	
Succinate (disodium salt)	162.1	
Sucrose	342.3	
Sulfuric Acid	98.1	18
N, N, N', N'-Tetramethylethylenediamine (TEMED)	116.2	
TES	229.25	
Trichloroacetic Acid (TCA)	163.4	
Tricine	179.2	
Tris	121.1	
Tween 20	1228	
Urea	60.1	
Zwittergent 3-14	363.6	

Note: An x% solution of compound A contains x grams of compound A in 100ml of solvent.

Appendix 2

Molecular Weights and Isoelectric Points of Selected Proteins

Protein	Molecular Weight	IEP
Cytochrome c	11,700	
Ribonuclease	13,700	
Lysozyme	14,300	
*Hemoglobin	15,500	
Myoglobin	17,200	
*β-Lactoglobulin	18,400	
Papain	23,000	8.75
Carbonic Anhydrase	29,000	
Carboxypeptidase	34,600	6.0
Pepsin	35,000	
*Glyceraldehyde-3-Phosphate Dehydrogenase	36,000	
*Lactate Dehydrogenase	36,000	
Tropomyosin	36,000	
*Alcohol Dehydrogenase (Yeast)	37,000	
*Aldolase	40,000	6.1
*Alcohol Dehydrogenase (Liver)	41,000	
*Enolase	41,000	
Ovalbumin	43,000	4.8
*Fumarase	49,000	
Leucine Aminopeptidase	53,000	
*Glutamate Dehydrogenase	53,000	
*Pyruvate Kinase	57,000	
*Catalase	60,000	5.8
Serum Albumin	68,000	
*Phosphorylase a	94,000	
*β-Galactosidase	130,000	
*Myosin Heavy Chain	220,000	

* exists as oligomer in native state

from Practical Handbook of Biochemistry and Molecular Biology. 1989. G.D. Fasman, ed. 601 pages. CRC Press, Inc., Boca Raton, Florida.

Appendix 3

Ammonium Sulfate Precipitation Table

Grams of Ammonium Sulfate to Add to a 1 Liter Solution

Starting Concentration	Final Concentration: 5%	10%	15%	20%	25%	30%	35%	40%	45%	50%
0%	27	55	84	113	144	176	208	242	277	314
5%		27	56	85	115	146	179	212	246	282
10%			28	57	86	117	149	182	216	251
15%				28	58	88	119	151	185	219
20%					29	59	89	121	154	188
25%						29	60	91	123	157
30%							30	61	92	126
35%								30	62	94
40%									31	63
45%										31

Starting Concentration	Final Concentration: 55%	60%	65%	70%	75%	80%	85%	90%	95%	100%
0%	351	390	430	472	516	561	608	657	708	761
5%	319	357	397	439	481	526	572	621	671	723
10%	287	325	364	405	447	491	537	584	634	685
15%	255	292	331	371	413	456	501	548	596	647
20%	223	260	298	337	378	421	465	511	559	609
25%	191	227	265	304	344	386	429	475	522	571
30%	160	195	232	270	309	351	393	438	485	533
35%	128	163	199	236	275	316	358	402	447	495
40%	96	130	166	202	241	281	322	365	410	457
45%	64	97	132	169	206	245	286	329	373	419
50%	32	65	99	135	172	210	250	292	335	381
55%		33	66	101	138	175	215	256	298	343
60%			33	67	103	140	179	219	261	305
65%				34	69	105	143	183	224	266
70%					34	70	107	146	186	228
75%						35	72	110	149	190
80%							36	73	112	152
85%								37	75	114
90%									37	76
95%										38

adapted from Protein Purification: Principles and Practice. 1982. R.K. Scopes. 282 pages. Springer-Verlag, New York.

Appendix 4

Spectrophotometer Linearity

In general, spectrophotometers are limited by stray light in their range of absorbances which can be read accurately. 0.1% stray light means that a solution with OD=3.0 (0.1% transmittance) will appear to have 0.2% transmittance or OD=2.7. Higher concentrations will never exceed an apparent OD of 3.0. It is useful to test your spectrophotometer in order to ascertain its range of linearity. Prepare a 0.5% solution of BSA and a series of dilutions to be measured. Plot the OD_{280} versus the concentration. The results will be linear up to the break, after which the OD is always below the extrapolated value (dashed line):

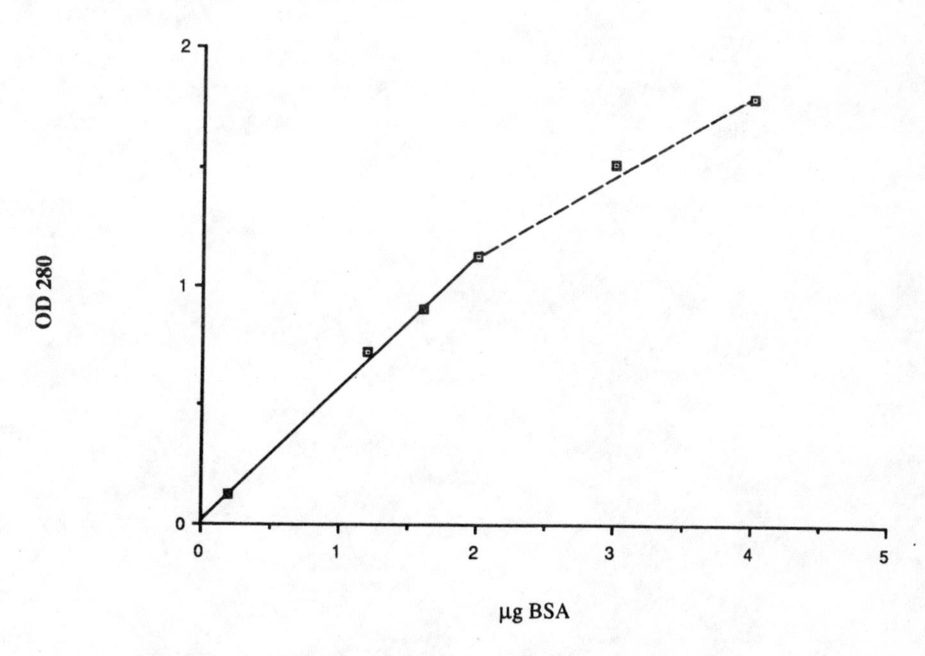

Figure A.1. Test of spectrophotometer linearity.

Conclusions: To be on the safe side, never exceed OD 1 at any wavelength relative to an absorbance by water of 0.

Keep slit width setting to a minimum, since wider slit widths may also limit the linearity range.

Appendix 5

Suppliers and Addresses

Aldrich Chemical Company, Inc.
940 West Saint Paul Ave.
Milwaukee, WI 53233
Tel. (414) 273-3850
(800) 558-9160

Amicon
24 Cherry Hill Drive
Danvers, MA 01923
Tel. (617) 777-3622
(800) 343-1397

Baxter Scientific Products
1430 Waukegan Road
McGaw Park, IL 60085
Tel. (312) 689-8410
(800) 633-7369

Bethesda Research Laboratories
P.O. Box 6009
Gaithersburg, MD 20877
Tel. (301) 840-8000
(800) 638-8992

Bio-Rad Laboratories
1414 Harbour Way, S.
Richmond, CA 94804
Tel. (415) 232-7000
(800) 4-BIORAD

Biospec Products
P.O. Box 722
Bartlesville, OK 74005
Tel. (918) 333-2166

B. Braun Instruments
824 Twelfth Ave.
Bethlehem, PA 18018
Tel. (215) 868-0300
(800) 258-9000

BDH Chemicals, Ltd.
Broom Road
Poole, Dorset
BH12 4NN UK

Calbiochem (Behring Diagnostics)
P.O. Box 12087
San Diego, CA 92112
Tel. (619) 450-9600
(800) 854-9256

CUNO, Inc.
400 Research Pky.
Meriden, CT 06450
Tel. (203) 237-5541
(800) 231-2259

Eastman Kodak Co.
343 State St., Building 701
Rochester, NY 14650
Tel. (716) 458-4014
(800) 225-5352

Fisher Scientific
50 Fadem Road
Springfield, NJ 07081
Tel. (201) 467-6400

Fluka Chem. Corp.
980 S. Second St.
Ronkonkoma, NY 11779
Tel. (516) 467-0980
(800) FLUKA-US

Gaulin Corporation
44 Garden Street
Everett, MA 02149

Hamilton Bonaduz AG
case postale 26
CH-7402 Bonaduz
Switzerland
Tel. 041 081 37 14 33

Hamilton Instruments
P.O. Box 100030
Reno, NV 89520
Tel. (702) 786-7077

Hoefer Scientific Instruments
P.O. Box 77387
654 Minnesota St.
San Francisco, CA 94107
Tel. (415) 282-2307
(800) 227-4750

Hunt Mfg. Co. (Speedball India Ink)
Statesville, NC

Kontron Instruments Inc.
9 Plymouth St.
Everett, MA 02149
Tel. (617) 389-6400
(800) 343-3297

Merck
5 Skyline Drive
Hawthorne, NY 10532
Tel. (914) 592-4660

Millipore Corp.
80 Ashby Rd.
Bedford, MA 01730
Tel. (617) 275-9200
(800) 225-1380

New Brunswick Scientific Co., Inc.
44 Talmadge Road
Edison, NJ 08818
Tel. (201) 287-1200
(800) 631-5417

New England Nuclear
549 Albany Street
Boston, MA 02118
Tel. (617) 350-9153
(800) 551-2121

Pellikan AG (Pelikan India Ink)
D-3000
Hanover 1
West Germany

Pharmacia LKB Biotechnology AB
S-75182
Uppsala, Sweden
Tel. 011 46 18 16 3000

800 Centennial Ave.
Piscataway, NJ 08854
Tel. (201) 457-8000
(800) 526-3618

Pierce Chemical Co.
P.O. Box 117
Rockford, IL 61105
Tel. (815) 968-0747
(800) 8-PIERCE

Polysciences, Inc.
400 Valley Rd.
Warrington, PA 18976
Tel. (215) 343-6484
(800) 523-2575

Sartorius Filters, Inc.
30940 San Clemente St.
Hayward, CA 94544
Tel. (415) 487-8220
(800) 227-2842

Schleicher & Schuell, Inc.
10 Optical Ave.
Keene, NH 03431
Tel. (603) 352-3810
(800) 245-4024

Serva Fine Biochemicals Inc.
200 Shames Drive
Westbury, NY 11590
Tel. (516) 333-1575
(800) 645-3412

Sigma Chemical Co.
P.O. Box 14508
St. Louis, MO 63178
Tel. (314) 771-5750
(800) 325-3010

Spectrum Medical Industries, Inc.
60916 Terminal Annex
Los Angeles, CA 90054
Tel. (213) 650-2100
(800) 634-3300

Thomas Scientific
99 High Hill Road
P.O. Box 99
Swedesboro, NJ 08085
Tel. (609) 467-2000
(800) 524-0027

United States Biochemical Corp.
P.O. Box 22400
Cleveland, OH 44122
Tel. (216) 765-5000
(800) 321-9322

VWR Scientific
P.O. Box 7900
San Francisco, CA 94120
Tel. (415) 467-6202
(800) 257-8407

Whatman, Inc.
9 Bridewell Pl.
Clifton, NJ 07014
Tel. (201) 773-5800
(800) 631-7290

INDEX